Dynamic Oligopolies with Time Delays

Akio Matsumoto · Ferenc Szidarovszky

Dynamic Oligopolies with Time Delays

Institute of Economic Research Chuo University
IERCU
中央大学経済研究所
1964

Springer

Akio Matsumoto
Department of Economics
Chuo University
Hachioji, Tokyo, Japan

Ferenc Szidarovszky
Department of Mathematics
Corvinus University
Budapest, Hungary

ISBN 978-981-13-1785-9 ISBN 978-981-13-1786-6 (eBook)
https://doi.org/10.1007/978-981-13-1786-6

Library of Congress Control Number: 2018958494

This Springer imprint is published by the registered company Springer Nature Singapore Pte Ltd.
The registered company address is: 152 Beach Road, #21-01/04 Gateway East, Singapore 189721, Singapore

Preface

The authors of this book spent several decades of performing research in different areas of mathematical economics and especially in game theory until the mid-1990s when they met in a conference in Odense, Denmark. Since then, they work together in examining different dynamic economic models with special attention to different variants of oligopolies. This cooperation resulted in many conference presentations, over 75 research papers, an introductory book on game theory and a jointly edited volume of selected articles in economic dynamics. They meet at least once every year in Tokyo and either in Tucson (Arizona) or in Budapest.

The objective of this book is to give an overview of recent developments in the asymptotic analysis of different versions of dynamic oligopolies with special attention to the presence of delayed information. This book can be considered as a continuation of some earlier works on oligopolies. The monograph of Okuguchi (1976) summarizes the earliest developments up to 1975. Their multiproduct generalizations are presented in Okuguchi and Szidarovszky (1999). The existence and uniqueness of the equilibrium were the main focus in addition to linear dynamic extensions. In the case of linear models, local asymptotical stability implies global stability making the asymptotical analysis relatively simple. Nonlinear oligopolies are introduced and analyzed in Bischi et al. (2010); however, no time delays are assumed in the models. The presence of delayed information is a realistic assumption in real economies, since data collection, determination of acceptable and optimal decisions and their implementations need time. This book is a significant addition to the earlier works by investigating the effects of delayed information on the asymptotical behavior of the state trajectories of the associated dynamic systems.

The methodology developed and used in this book can be useful in other fields of economic dynamics, especially when instantaneous information is not available and therefore the decision makers have only delayed data. The methodology and the actual research results of this book might have practical values when decision makers base their actual decisions on sophisticated mathematical modeling and methodology.

The authors strongly recommend this book to undergraduate and graduate students, researchers, since the mathematical methodology and therefore the results reported in this book can be generalized and modified accordingly to other similar and even in more complex cases. This book also can help members of managerial teams responsible to reach reasonable and profitable decisions under realistic conditions. The authors are thankful to the large number of reviewers of the earlier papers of the authors as well as to the participants of many conferences on nonlinear dynamics for their useful comments which helped to make the quality of the published papers and therefore that of the corresponding sections of this book better. The authors are also thankful to the editorial staff of the Tokyo office of Springer for their assistance in preparing the final version of this book. Further, the authors appreciate financial supports: Graduate School of Economics of Chuo University with the MEXT-supported Program for the Strategic Research Foundation at Private Universities 2013–2017, the Japan Society for the Promotion of Science (Grant-in-Aid for Scientific Research (C), 24530202, 25380238, 26380316 and 16K03556) and Joint Research Grant of Chuo University.

Hachioji, Japan — Akio Matsumoto
Budapest, Hungary — Ferenc Szidarovszky

Contents

Chapter 1
Introduction

Game theory is one of the most important fields of mathematical economics. It is the basis of modeling situations when more than one decision makers (called players) deal with an issue and the decision of each of them affects the outcome of all of them. If they select their decisions simultaneously and independently of each other, then the game is considered noncooperative and the Nash equilibrium provides the usual solution. The players can exchange information and take the interest of others into consideration, when the game is partially or completely cooperative. In the case of leader–follower games, the players make their decisions sequentially and at each time the decisions are made known to all players. Therefore, each player's decision depends on the most recent choices of the others. There are also several variants and extensions of the different solution concepts including Stackelberg equilibria, subgame perfect Nash equilibria, Bayesian and perfect Bayesian equilibria; however, in this book we concentrate only on the classical Nash equilibrium concept and cases of partial cooperation.

In the literature on game theory, the oligopoly models play a very special role. A classical oligopoly model describes a situation when a relatively small number of firms produce the same or related products which are sold in the same market, or offer the same or related services. The market is assumed to be homogeneous with a deterministic relation between the supplied quantities and market prices. The most simple model assumes one identical product which is called oligopoly without product differentiation, and if the firms produce different products and the price of each product is a function of the quantities of all products, then the model is called oligopoly with product differentiation. There are many variants of this model including labor managed, market sharing and rent-seeking games, multiproduct oligopolies, group equilibrium problems and oligopsonies. They can be analyzed similarly to the cases shown in this book. Another distinction between the model types is the choice of the decision variables of the firms. In the case of Cournot oligopolies, the produced and sold quantities are the choices of the firms and the unit prices are then determined by the inverse demand functions. In the case of Bertrand oligopolies, the firms

A. Matsumoto and F. Szidarovszky, *Dynamic Oligopolies with Time Delays*,
https://doi.org/10.1007/978-981-13-1786-6_1

select the prices for their products and then the market demand functions determine their corresponding production quantities. Another distinction is the choice of the time scale. In the discrete case, only discrete time periods are considered, which are assumed to be the nonnegative integers for the sake of mathematical simplicity. In the continuous case, the time periods can take any nonnegative real values.

In the huge literature on oligopoly theory, first the existence and uniqueness of the equilibrium was the main focus, and then dynamic extensions became the main research area. In formulating dynamic extensions, usually two types of dynamisms are introduced: simultaneous and sequential move models. In this book, we will concentrate on the first type. At each time period, the firms adjust their output levels or prices by one of the following concepts. In the case of best response dynamics, each firm determines its best response based on its belief about the choices of the competitors and then adjust its output level or price into the direction toward its best response. In the case of gradient dynamics, the firms compute their marginal profits and adjust their decision variables proportionally to the marginal profits. In this book, examples will be shown for both types of dynamic models. In the early stages, only linear models were considered, when the asymptotic properties could be easily examined and local asymptotical stability implied global stability. The next stage was the introduction of nonlinearities into oligopolies by introducing nonlinear price and cost functions as well as requiring bounded feasible decision spaces such as defining capacity limits in Cournot oligopolies. The local asymptotic behavior of the equilibria could be examined by selecting a Lyapunov function or by local linearization. Global analysis can be performed by using an appropriate Lyapunov function, computer simulation or in the discrete time scales by the theory of critical curves. All these models assumed that at each time period, each firm had instantaneous information about the output volumes and prices of all firms; however, in real economies this simplifying assumption had to be modified. Before a firm decides about its optimal output level or price, it has to collect all necessary information, determine its optimal choice and implement it. All these need time leading to information lags, which are taken into account in the mathematical models as delayed dynamics. If the delay is due to institutional or contractual circumstances, then fixed delays are assumed. The lengths of the delays are considered fixed as any positive value in the continuous case and a positive integer in the discrete case. In the continuous case, the characteristic equation is a mixed polynomial–exponential equation. In the discrete case, the introduction of additional shifting variables can reduce the problem to a higher dimensional model without delay. If the delays are uncertain due to the different delays of a large number of firms, or the firms want to react to average past information rather than to sudden market changes, then continuously distributed delays are the appropriated approach. In the continuous case, Volterra-type integro-differential equations are used as mathematical models. In the discrete case, models with geometric delays describe the distributed delays with appropriate weighting sequences.

The rest of this book consists of five chapters and three appendices. Chapter 2 discusses the most simple case of only one firm, called monopoly. Chapter 3 is devoted to the cases of two firms, which are called duopolies, and Chap. 4 discusses the most general case of n-firm oligopolies. Most models of monopolies are single-dimensional dynamics, which can be examined with relatively simple mathematical tools. Duopolies mostly result in two-dimensional models, which are more complicated than their single-dimensional counterparts, but the degrees of the characteristic functions are low allowing us to apply simple stability conditions. Even in the linear case, there is a significant difference between duopolies and n -firm ($n > 2$) oligopolies with respect to the asymptotic behavior of their equilibria. The n-firm oligopolies lead to high-degree characteristic equations, and there are no simple sufficient and necessary stability conditions, and in some cases simple sufficient conditions are derived or special cases, like symmetry, are considered. Linear or quadratic cost functions are assumed, but two kinds of price functions are selected: linear and hyperbolic. In addition, both Cournot and Bertrand cases are described in these chapters, which are divided into three subsections: discrete models, continuous models with fixed delays and continuous models with continuously distributed delays. The case of discrete geometric delay is introduced for monopolies as an illustration. In all subsections, Cournot and Bertrand oligopolies are discussed without delay and then with delays. Also both linear and hyperbolic price functions are considered. Chapter 5 is devoted to some learning models when the firms want to learn the value of certain parameter of the linear price function. Three particular models are introduced, and the asymptotic stability of the associated dynamic systems is analyzed which provides successful learning. Both discrete and continuous time scales as well as fixed and continuously distributed delays are considered. This chapter has two sections: monopolies and oligopolies. The last Chap. 6 first introduces several profit-sharing schemes and joint ventures of the firms which are mathematically equivalent with partial cooperation of the firms. For the sake of simplicity, only linear Cournot oligopolies are considered in both discrete and continuous time scales. Then it is examined how the dynamic behavior of the equilibrium changes if a group of firms cooperate with each other, while the others do not. The cartelizing group can be punished if the authorities notice their antitrust behavior. The Herfindahl–Hirschman index is used to detect this illegal activity, which is also monitored by the cartelizing firms. A threshold is selected which is used by them to approach the noncooperative or the cooperative equilibrium leading to an interesting flip-flop dynamics. This chapter is therefore divided into two sections: partial cooperation and cartelizing groups with antitrust threshold. The last section of each chapter contains supplementary notes and discussions providing the sources of the material of the different sections and offers additional references for further reading. The basic mathematical results about stability of delayed systems are placed in the three appendices. These results are used many times repeatedly in the book and to separate them from the main body of the book seemed the best way to keep the continuity of the discussions of the different chapters. Appendix A discusses fixed delays in continuous time scales, the cases on one, two and even three delays are considered. Continuously distributed delays are the subject of Appendix B which includes the famous Routh–Hurwitz stability criteria. The stability conditions of discrete and continuous systems are compared in Appendix C, which includes some well-known stability conditions for discrete systems based on the coefficients of the characteristic polynomials. There are still many unanswered questions concerning different model variants, types of solution concepts, different forms of the cost and price functions and delays serving as potential research areas for the interested readers. The authors of this book hope that many talented researchers will turn to this area to enrich this very interesting and exciting field.

Chapter 2
Dynamic Monopolies

Monopoly models describe an economic situation where a single firm produces a homogeneous product or offers a service to a homogeneous market. Its objective is to maximize profit.

Let q denote the production output of the firm, $p(q)$ be the unit price on the market and $C(q)$ be the production cost. Function $p(q)$ strictly decreases in q, and $C(q)$ strictly increases in q. The profit of the firm can be given as the difference of its revenue and cost,

$$\pi(q) = qp(q) - C(q). \tag{2.1}$$

This model also includes the case when there are competitors in the market, but their total production level Q is known by the firm and does not change in time. Then, the price function is $p(q + Q)$; however with fixed value of Q, it really depends on only q.

For most firms, there is a capacity limit L meaning that the feasible set of the firm's output is the closed interval $[0, L]$. Even if there is no capacity limit, the price becomes zero with sufficiently large value of the output, so the firm will not produce more than this threshold. In either case, the feasible output set is a finite closed interval, and if functions $p(q)$ and $C(q)$ are continuous, there is a finite nonnegative profit-maximizing output level. It will be assumed that the firm is *boundedly rational* meaning that it does not have complete knowledge about the price function and experiences time delays in information and implementation.

In the mathematical economics literature, two types of delays are considered. If the delay is due to institutional or contractional circumstances, then fixed delays are the appropriate approaches (Bellman and Cooke 1963). However, if the delay is uncertain, or the firm does not want to react to sudden market changes, then instead it considers and uses an average of past information. Continuously, distributed delays are then assumed (Cushing 1977).

A. Matsumoto and F. Szidarovszky, *Dynamic Oligopolies with Time Delays*,
https://doi.org/10.1007/978-981-13-1786-6_2

In this chapter, we will discuss a group of dynamic monopolies with both continuously distributed and fixed delays. We will also consider discrete and continuous time scales.

2.1 Discrete Time Models

For the sake of simplicity, assume linear price function and quadratic cost:

$$p(q) = a - bq$$

and

$$C(q) = cq^2,$$

where $a,\ b$ and c are positive constants. It is assumed that the monopolistic firm has only partial information about the price function and its decision is based on an average of past output values, q^e. The marginal profit of the firm is given as

$$\frac{\partial \pi}{\partial q} = a - 2(b + c)q^e. \tag{2.2}$$

The right hand side is the average past marginal profit because of its linearity. The firm adjusts its output in such a way that the output growth rate is proportional to the average marginal profit leading to the dynamic equation

$$\frac{q(t+1) - q(t)}{q(t)} = \alpha \frac{\partial \pi \left(q^e(t+1)\right)}{\partial q}$$

where $q^e(t+1)$ can be imagined as the firm's output expectation in period $t+1$. This equation is called *gradient dynamics* and can be written as

$$q(t+1) = q(t) + \alpha q(t) \left[a - 2(b + c)q^e(t+1)\right]. \tag{2.3}$$

The dynamics of this system depends on the selection of the average past demands.

First, we consider the case of a special *one-step delay* when $q^e(t+1)$ is selected as the actual output of the firm in the previous time period,

$$q^e(t+1) = q(t), \tag{2.4}$$

in which case, Eq. (2.3) becomes

$$q(t+1) = q(t) + \alpha q(t) \left[a - 2(b + c)q(t)\right]. \tag{2.5}$$

This system has two steady states. One is $\bar{q} = 0$, and the other is the positive solution of equation

$$\bar{q} = \bar{q} + \alpha\bar{q}\,[a - 2(b + c))\bar{q}]$$

which is

$$\bar{q} = \frac{a}{2(b + c)}. \tag{2.6}$$

We will call this positive steady state as the *monopoly equilibrium*.

It is well known that system (2.5) is locally asymptotically stable if the absolute value of the derivative of the right hand side of Eq. (2.5) with respect to $q(t)$ at $q(t) = \bar{q}$ is less than unity. Since

$$\frac{\partial q(t+1)}{\partial q(t)} = 1 + \alpha a - 4\alpha(b + c)q(t)$$

and its value at $q(t) = \bar{q}$ becomes

$$1 + \alpha a - 4\alpha(b + c)\frac{a}{2(b + c)} = 1 - \alpha a,$$

the stability condition is

$$|1 - \alpha a| < 1$$

or

$$-1 < 1 - \alpha a < 1.$$

Since both α and a are positive, this condition is equivalent to

$$\alpha a < \alpha a^* = 2. \tag{2.7}$$

We can next show an interesting relation between system (2.5) and the logistic map. Introduce the new variable,

$$x(t) = \frac{q(t)\,[2\alpha(b + c)]}{1 + \alpha a},$$

and then Eq. (2.5) becomes

$$x(t + 1) = (1 + \alpha a)x(t)\,(1 - x(t)) \tag{2.8}$$

with confinement condition $\alpha a \leq 3$. If $\alpha a < 2$, then the system is locally asymptotically stable, and if the value of αa increases from 2 to 3, then the steady state is destabilized by a *period doubling (PD) bifurcation* (see, for example, Martelli 1999) in which the stability is replaced by a 2-periodic cycle, then further replaced by a 4-periodic cycle and so on. So, we have the following result.

Proposition 2.1 *In the case of one-step delay* (2.4), *the monopoly equilibrium $\bar{q}$ is locally asymptotically stable if $\alpha a < \alpha a^* = 2$, loses stability at $\alpha a = 2$ and undergoes a PD cascade to chaos, while the value of αa increases from* 2 *to* 3.

Next, we consider a *two-step delay*, when

$$q^e(t+1) = \omega_0 q(t) + \omega_1 q(t-1) \tag{2.9}$$

with $\omega_1 = 1 - \omega_0$, when system (2.3) has the following form:

$$q(t+1) = q(t) + \alpha q(t)\left[a - 2(b+c)\left(\omega_0 q(t) + (1-\omega_0)q(t-1)\right)\right] \tag{2.10}$$

which is equivalent to a two-dimensional first-order difference equation,

$$\begin{aligned} x(t+1) &= q(t), \\ q(t+1) &= q(t) + \alpha q(t)\left[a - 2(b+c)\left(\omega_0 q(t) + (1-\omega_0)x(t)\right)\right]. \end{aligned} \tag{2.11}$$

The steady state of this system is clearly

$$x = q = \bar{q} = \frac{a}{2(b+c)}.$$

Simple differentiation shows that

$$\frac{\partial x(t+1)}{\partial x(t)} = 0,$$

$$\frac{\partial x(t+1)}{\partial q(t)} = 1$$

and

$$\frac{\partial q(t+1)}{\partial x(t)} = \alpha q(t)\left[-2(b+c)(1-\omega_0)\right]$$

which becomes

$$\alpha\frac{a}{2(b+c)}\left[-2(b+c)(1-\omega_0)\right] = -\alpha a(1-\omega_0)$$

at the steady state. Similarly,

$$\frac{\partial q(t+1)}{\partial q(t)} = 1 + \alpha\left[a - 2(b+c)\left(\omega_0 q(t) + (1-\omega_0)x(t)\right)\right] - 2\alpha(b+c)\omega_0 q(t) = 1 - \alpha a \omega_0$$

at the steady state. So, the Jacobian of the system has the form

$$\mathbf{J} = \begin{pmatrix} 0 & 1 \\ -\alpha a(1-\omega_0) & 1-\alpha a\omega_0 \end{pmatrix} \tag{2.12}$$

with characteristic polynomial

$$\begin{aligned} \varphi(\lambda) &= -\lambda\,(1-\alpha a\omega_0-\lambda)+\alpha a(1-\omega_0) \\ &= \lambda^2+a_1\lambda+a_2 \end{aligned}$$

with

$$a_1 = -\,(1-\alpha a\omega_0)\ \text{ and } a_2 = \alpha a(1-\omega_0).$$

It is well known that system (2.11) is locally asymptotically stable if all eigenvalues are inside the unit cycle, which is the case if and only if

$$a_2 < 1$$

and

$$\pm a_1 + a_2 + 1 > 0.$$

(see Appendix C for details). In our case, these conditions can be written as follows:

$$\begin{aligned} & a_1+a_1+1 = \alpha a > 0, \\ & -a_1+a_2+1-2-\alpha a(2\omega_0-1) > 0, \\ & 1-a_2 = 1-\alpha a(1-\omega_0) > 0. \end{aligned} \tag{2.13}$$

The first condition always holds, and the second and third conditions are satisfied only under a certain assumption about model parameters. For example, they hold if

$$\frac{-1+\alpha a}{\alpha a} < \omega_0 < \frac{2+\alpha a}{2\alpha a}. \tag{2.14}$$

We next turn our attention to the special case of $\omega_0 = 0$ and $\omega_1 = 1$. The other boundary case of $\omega_0 = 1$ and $\omega_1 = 0$ was already considered in assumption (2.4). Notice first that condition (2.14) becomes $-1 + a\alpha \leq 0$, so the stability threshold is

$$a\alpha^{**} = 1.$$

If $a\alpha < 1$, then the system is locally asymptotically stable. As the value of $a\alpha$ increases, the stability is first replaced with a periodic cycle which is then replaced with a quasiperiodic cycle. This type of stability change is known as a *Neimark–Sacker (NS) bifurcation* (see, for example, Kuznetsov and Sacker 2008). Notice that the threshold $a\alpha = 1$ is stronger than the threshold $a\alpha = 2$ in the case of assumption

(2.4); therefore, the special two-delay model with $\omega_0 = 0$ and $\omega_1 = 1$ has stronger destabilizing effect than the model with $\omega_0 = 1$ and $\omega_1 = 0$. We have now the following result.

Proposition 2.2 *Under assumption* $\omega_0 = 0$ *and* $\omega_1 = 1$, *the monopoly equilibrium is locally asymptotically stable if* $a\alpha < a\alpha^{**} = 1$ *and loses stability as* $a\alpha = 1$ *and as* $a\alpha > 1$; *it is locally unstable through a NS bifurcation.*

The general two-step delay with $\omega_0 > 0$ and $\omega_1 > 0$ is now examined. The monopoly equilibrium is locally asymptotically stable if conditions (2.13) hold. The first condition always holds, and the second also holds if $\omega_0 \leq 1/2$ for all values of $a\alpha$. If $\omega_0 > 1/2$, then the *PD* boundary is given as

$$a\alpha^I_{PD} = \frac{2}{2\omega_0 - 1} \; (> 2) \, . \tag{2.15}$$

If the monopoly equilibrium crosses this boundary, then one of the eigenvalues becomes -1 and then *PD* bifurcation occurs. From the third condition of (2.13), we see that the *NS* boundary is defined as

$$a\alpha^I_{NS} = \frac{1}{1 - \omega_0} \; (>1) \, . \tag{2.16}$$

On this curve, there are two pure complex roots with zero real parts. If the value of $a^*\alpha$ crosses this boundary, then *NS* bifurcation occurs. Combining (2.15) and (2.16), it is clear that the partition curve is defined as

$$a\alpha = \min \left\{ a\alpha^I_{PD}, \, a\alpha^I_{NS} \right\} \tag{2.17}$$

which divides the $(\omega_0, a\alpha)$ plane into two regions: The monopoly equilibrium is locally asymptotically stable under this curve and locally unstable above it. Figure 2.1 shows a two-parameter bifurcation diagram for $0 < \omega_0 < 1$ and $0 < a\alpha < 4$. Notice that the upward sloping curve is the *NS* boundary and the downward sloping curve is the *PD* boundary. The two curves intersect at $\omega_0 = 3/4$. The monopoly equilibrium is locally asymptotically stable in the red region under the two curves. The different cycle periods up to 16 are indicated by different colors. If the period is longer than 16 or aperiodic cycles are born, then the corresponding region is shown in gray color. The white region indicates that solution becomes infeasible. We can summarize the above discussion as follows.

Proposition 2.3 *In case of* $\omega_0 > 0$ *and* $\omega_1 > 0$, *the plane* $(\omega_0, a\alpha)$ *is divided into two regions by the partition curve*

$$a^*\alpha = \min \left\{ a\alpha^I_{PD}, \, a\alpha^I_{NS} \right\},$$

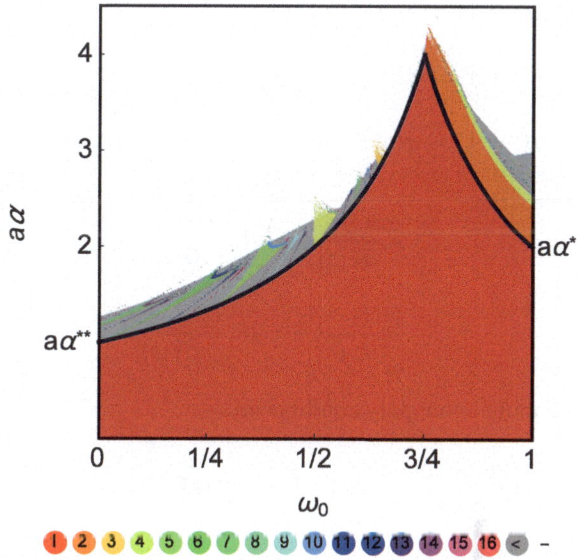

Fig. 2.1 Bifurcation diagram with a general two-step delay

where stability is lost. If stability is lost for $0 < \omega_0 < 3/4$ *at* $a\alpha = a\alpha_{NS}^I$*, then NS bifurcation occurs, and if stability is lost for* $3/4 < \omega_0 < 1$ *at* $a\alpha = a\alpha_{PD}^I$*, then PD bifurcation occurs.*

We can make the following observations.

(i) The general two-step delay has a less stable monopoly equilibrium than in the case of the one-step delay if $0 < \omega_0 < 1/2$, since $a\alpha_{NS}^I < a\alpha^* = 2$.
(ii) If $1/2 < \omega_0 < 1$, then $\min\left\{a\alpha_{PD}^I,\ a\alpha_{NS}^I\right\} > a\alpha^* = 2$; therefore, the general two-step delay leads to more stable monopoly equilibrium than the one-step delay.

Three-step delay will be next discussed, when

$$q^e(t+1) = \omega_0 q(t) + \omega_1 q(t-1) + \omega_2 q(t-2) \tag{2.18}$$

where $\omega_0,\ \omega_1,\ \omega_2 \geq 0$ and $\omega_0 + \omega_1 + \omega_2 = 1$.

Assume first that $\omega_0,\ \omega_1,\ \omega_2$ are all positive. Then, Eq. (2.3) becomes

$$q(t+1) = q(t) + \alpha q(t)\left[a - 2(b+c)\left(\omega_0 q(t) + \omega_1 q(t-1) + \omega_2 q(t-2)\right)\right] \tag{2.19}$$

which can be rewritten as a three-dimensional first-order difference equation,

$$
\begin{aligned}
x(t+1) &= y(t), \\
y(t+1) &= q(t), \\
q(t+1) &= q(t) + \alpha q(t)\left[a - 2(b+c)\left(\omega_0 q(t) + \omega_1 y(t) + \omega_2 x(t)\right)\right].
\end{aligned}
\tag{2.20}
$$

Notice that

$$
\frac{\partial x(t+1)}{\partial x(t)} = 0, \quad \frac{\partial x(t+1)}{\partial y(t)} = 1, \quad \frac{\partial x(t+1)}{\partial q(t)} = 0,
$$

$$
\frac{\partial y(t+1)}{\partial x(t)} = 0, \quad \frac{\partial y(t+1)}{\partial y(t)} = 0, \quad \frac{\partial y(t+1)}{\partial q(t)} = 1,
$$

and at the monopoly equilibrium

$$
\frac{\partial q(t+1)}{\partial x(t)} = \alpha\bar{q}\left(-2(b+c)\right)(1-\omega_0-\omega_1) = -\alpha a(1-\omega_0-\omega_1),
$$

$$
\frac{\partial q(t+1)}{\partial y(t)} = \alpha\bar{q}\left(-2(b+c)\right)\omega_1 = -\alpha a\omega_1,
$$

$$
\frac{\partial q(t+1)}{\partial q(t)} = 1 + \alpha\left[a - 2(b+c)\bar{q}\right] + \alpha\bar{q}\left(-2(b+c)\omega_0\right) = 1 - \alpha a\omega_0.
$$

Therefore, the Jacobian of system (2.20) has the form:

$$
\boldsymbol{J} = \begin{pmatrix} 0 & 1 & 0 \\ 0 & 0 & 1 \\ -\alpha a(1-\omega_0-\omega_1) & -\alpha a\omega_1 & 1-\alpha a\omega_0 \end{pmatrix}
\tag{2.21}
$$

with characteristic polynomial

$$
\lambda^3 + a_1\lambda^2 + a_2\lambda + a_3 = 0
$$

where the coefficients are as follows:

$$
\begin{aligned}
a_1 &= a\alpha\omega_0 - 1, \\
a_2 &= a\alpha\omega_1, \\
a_3 &= a\alpha(1-\omega_0-\omega_1).
\end{aligned}
$$

It is well known (see Appendix C for details) that all roots of this cubic polynomial are inside the unit circle if and only if

$$
\begin{aligned}
&1 + a_1 + a_2 + a_3 = a\alpha > 0, \\
&1 - a_1 + a_2 - a_3 = 2 - a\alpha(1 - 2\omega_1) > 0, \\
&1 - a_2 + a_1 a_3 - a_3^2 = \varphi(\omega_0, \omega_1, a\alpha) > 0, \\
&3 - a_2 = 3 - a\alpha\omega_1 > 0
\end{aligned}
\tag{2.22}
$$

where

$$\varphi(\omega_0, \omega_1, a\alpha) = 1 - a\alpha(1 - \omega_0) - (a\alpha)^2 \left[2\omega_0^2 - 3\omega_0(1 - \omega_1) + (1 - \omega_1)^2\right].$$

The first condition is always satisfied. The second always holds if $\omega_1 \geq 1/2$, and if $\omega_1 < 1/2$, then it provides the *PD* boundary

$$a\alpha < a\alpha_{PD}^{II} = \frac{2}{1 - 2\omega_1} \,(>2)\,. \tag{2.23}$$

The third condition is a quadratic inequality in $a\alpha$, so the boundary of $\varphi(\omega_0, \omega_1, a\alpha) = 0$ depends on the sign of the quadratic coefficient A, which can be factored as

$$A = 2\left(\omega_0 - (1 - \omega_1)\right)\left(\omega_0 - \frac{1}{2}(1 - \omega_1)\right).$$

This expression is zero if either

$$\omega_0 + \omega_1 - 1 = -\omega_2 = 0$$

or

$$\omega_0 = \frac{1}{2}(1 - \omega_1). \tag{2.24}$$

The first case of $\omega_2 = 0$ has been analyzed earlier, so we assume that the second case applies, in which case $\omega_0 < 1/2$ and the third condition of (2.22) becomes

$$a\alpha < \frac{1}{1 - \omega_0} = a\alpha_{NS}^{I}.$$

From (2.24), we have that $\omega_1 = 1 - 2\omega_0$ and if we substitute it into (2.23), we get an alternative form of the *PD* boundary,

$$a\alpha_{PD}^{II} = \frac{2}{4\omega_0 - 1}.$$

The last condition holds if

$$a\alpha < \frac{3}{\omega_1} = \frac{3}{1 - 2\omega_0}.$$

Notice next that

$$a\alpha_{PD}^{II} - a\alpha_{NS}^{I} = \frac{3\,(1-2\omega_0)}{(4\omega_0-1)\,(1-\omega_0)} > 0$$

as well as

$$\frac{3}{\omega_1} = \frac{3}{1-2\omega_0} > a\alpha_{NS}^{I} = \frac{1}{1-\omega_0} \text{ if } \omega_0 < \frac{1}{2}.$$

Therefore, $a\alpha < a\alpha_{NS}^{I}$ is the strongest condition implying the following result

Proposition 2.4 *Assume* $0 < \omega_1 < 1$ *and* $\omega_0 = \frac{1}{2}(1-\omega_1)$. *Then,* $a\alpha = a\alpha_{NS}^{I}$ *is the partition curve. The monopoly equilibrium is locally asymptotically stable below this curve, and above the curve it loses stability via a NS bifurcation for* $0 < \omega_0 < \frac{1}{2}$.

Next, we assume that the leading coefficient $\varphi(\omega_0, \omega_1, a\alpha)$ is nonzero. The root of this quadratic polynomial produces the two *NS* boundaries:

$$a\alpha_{(+)} = \frac{\omega_0 - 1 + \sqrt{D}}{2A} \tag{2.25}$$

and

$$a\alpha_{(-)} = \frac{\omega_0 - 1 - \sqrt{D}}{2A} \tag{2.26}$$

where the discriminant is

$$D = 5 + 9\omega_0^2 - 4\omega_1(2-\omega_1) + 2\omega_0(6\omega_1 - 7).$$

If $D < 0$, then there are no real roots. If $D = 0$, then the two roots are identical, and if $D > 0$, then the roots are different. Equation $D = 0$ gives two threshold values for ω_0,

$$\omega_0^{(+)} = \frac{7 - 6\omega_1 + 2\sqrt{1-3\omega_1}}{9} \tag{2.27}$$

and

$$\omega_0^{(-)} = \frac{7 - 6\omega_1 - 2\sqrt{1-3\omega_1}}{9}. \tag{2.28}$$

If $\omega_1 > 1/3$, then no solution exists, if $\omega_1 = 1/3$, then the two threshold values are equal, and there are two threshold values if $\omega_1 < 1/3$. In the second case,

$$\omega_0^{(+)} = \omega_0^{(-)} = \frac{5}{9},$$

and in the third case

$$\frac{1}{2} \le \omega_0^{(-)} < \frac{5}{9} < \omega_0^{(+)} < 1.$$

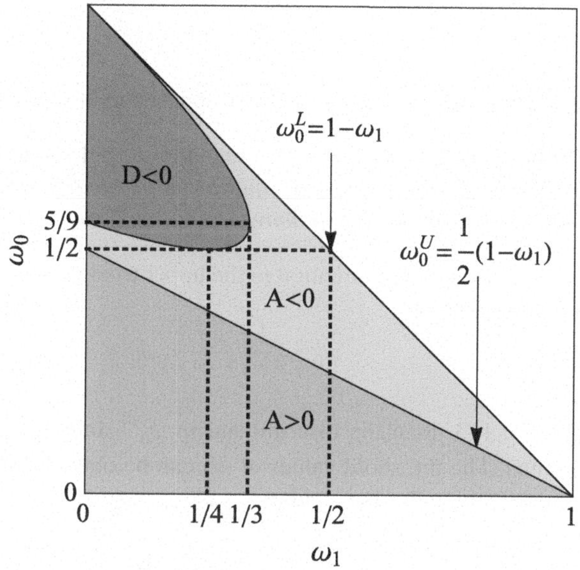

Fig. 2.2 Division of the (ω_1, ω_0) plane

Figure 2.2 shows the division of the (ω_1, ω_0) plane. The area above the line $\omega_0 = 1 - \omega_1$ is eliminated, since $\omega_0 + \omega_1 > 1$. The area under this line is further divided into two smaller triangles by the $\omega_0 = \frac{1}{2}(1 - \omega_1)$ line. Notice that $A < 0$ in the upper triangle shown with light gray and $A > 0$ in the lower triangle with medium gray. The dark gray region in the upper triangle with the half-oval shape shows when $D < 0$.

It can be proved that the lower part of the $D = 0$ curve, when $\omega_0 = \omega_0^{(-)}$, is downward sloping as $0 < \omega_1 < 1/4$, upward sloping as $1/4 < \omega_1 < 1/3$ and its minimal value is $1/2$ when $\omega_1 = 1/4$. The upper part of the $D = 0$ curve, when $\omega_0 = \omega_0^{(+)}$, is downward sloping and below the $\omega_0 = 1 - \omega_1$ line. For $\omega_1 > 1/3$, the discriminant is always positive, and for $\omega_1 > 1/2$, the second condition of (2.22) must hold.

We can now turn our attention to the stability of the monopoly equilibrium.

Consider first the lower triangle with $A > 0,\ D > 0$ and $\omega_0 < 1/2$. In examining the two *NS* boundaries (2.25) and (2.26), we need to check the signs of the numerators. Since

$$(1 - \omega_0)^2 - D = \left[\left(1 - \omega_0 + \sqrt{D}\right)\left(1 - \omega_0 - \sqrt{D}\right)\right]$$

$$= 8\,(1 - \omega_0 - \omega_1)\left(\omega_0 - \frac{1}{2}\,(1 - \omega_1)\right)$$

$$= -4A < 0,$$

the value of $a\alpha_{(-)}$ is negative and only the positive $a\alpha_{(+)}$ provides boundary. Furthermore, $a\alpha_{PD}^{II} > 2$ and it is easy to show that

$$a\alpha_{(+)} < a\alpha_{PD}^{II} \text{ and } a\alpha_{(+)} < 2 < \frac{3}{\omega_1}$$

implying that $a\alpha < a\alpha_{(+)}$ is the strongest condition, so we have the following result.

Proposition 2.5 *If $\omega_0 < \frac{1}{2}(1-\omega_1)$ and $\omega_1 < 1$, then the partition curve is given by $a\alpha = a\alpha_{(+)}$. The monopoly equilibrium is locally asymptotically stable below the curve, and stability is lost above the curve through a NS bifurcation.*

We turn next our attention to the upper triangle with $A < 0$. The feasible interval of ω_0 is given by

$$\omega_0^L = \frac{1-\omega_1}{2} < \omega_0 < 1-\omega_1 = \omega_0^U.$$

If $\omega_1 < 1/3$, then the discriminant in $\omega_0^{(+)}$ and $\omega_0^{(-)}$ is positive, so they are well defined. The threshold values of $a\alpha$ can be obtained by substituting $\omega_0^{(+)}$ and $\omega_0^{(-)}$, respectively, into $a\alpha_{(+)}$ and $a\alpha_{(-)}$:

$$a\alpha_m = \frac{9\left(1+3\omega_1+\sqrt{1-3\omega_1}\right)}{2+3\omega_1+9\omega_1^2+2\sqrt{1-3\omega_1}\,(1+3\omega_1)} \tag{2.29}$$

and

$$a\alpha_M = \frac{9\left(1+3\omega_1-\sqrt{1-3\omega_1}\right)}{2+3\omega_1+9\omega_1^2-2\sqrt{1-3\omega_1}\,(1+3\omega_1)}, \tag{2.30}$$

where

$$a\alpha_m - a\alpha_M = -\frac{2\sqrt{1-3\omega_1}}{\omega_1(1+\omega_1)} < 0.$$

It is also easy to show that

$$a\alpha_{(+)} = a\alpha_{(-)} = a\alpha_m \text{ at } \omega_0 = \omega_0^{(-)}$$

and

$$a\alpha_{(+)} = a\alpha_{(-)} = a\alpha_M \text{ at } \omega_0 = \omega_0^{(+)}.$$

The graphs of $a\alpha_m$, $a\alpha_M$, $a\alpha_{PD}^{II}$ and $3/\omega_1$ are illustrated in Fig. 2.3.

It is easy to prove that

$$a\alpha_{PD}^{II} < a\alpha_m < a\alpha_M \text{ if } \omega_1 < \frac{1}{4},$$

$$a\alpha_m < a\alpha_{PD}^{II} < a\alpha_M \text{ if } \frac{1}{4} < \omega_1 < \frac{5}{16}$$

and

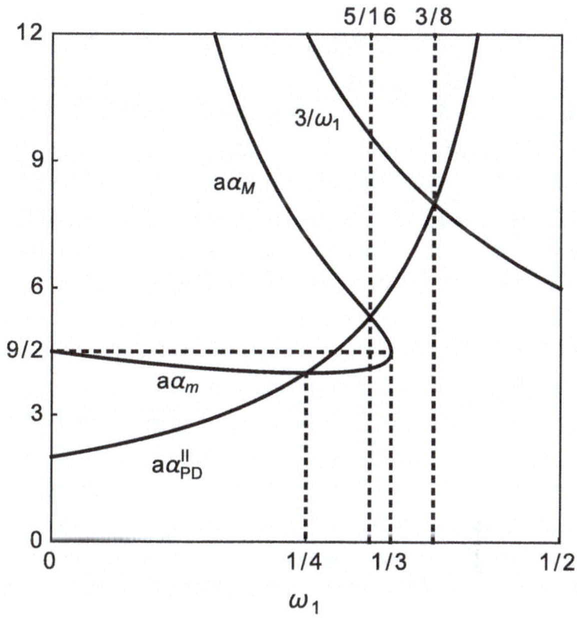

Fig. 2.3 Graphs of $a\alpha_m$, $a\alpha_M$, $a\alpha^{II}_{PD}$ and $3/\omega_1$

$$a\alpha_m < a\alpha_M < a\alpha^{II}_{PD} \text{ if } \frac{5}{16} < \omega_1 < \frac{1}{3}$$

as well as the $3/\omega_1$ curve is above those of $a\alpha_M, a\alpha_m$ and $a\alpha^{II}_{PD}$ if $\omega_1 \leq 1/3$. If $1/3 < \omega_1 < 3/8$, then $a\alpha^{II}_{PD}$ is below the $3/\omega_1$ curve, and if $\omega_1 > 3/8$, then it is above the $3/\omega_1$ curve. They intersect at $\omega_1 = 3/8$. Figure 2.4 shows three boundaries with three different values of ω_1. The $a\alpha = a\alpha_{(+)}$ curve is colored in red, the $a\alpha = a\alpha_{(-)}$ curve is blue, and the $a\alpha = a\alpha^{II}_{PD}$ curve is shown with dotted horizontal line. We can now check the stability condition (2.22). The third condition with positive discriminant holds in the (darker or lighter) gray regions as well as it also holds with negative

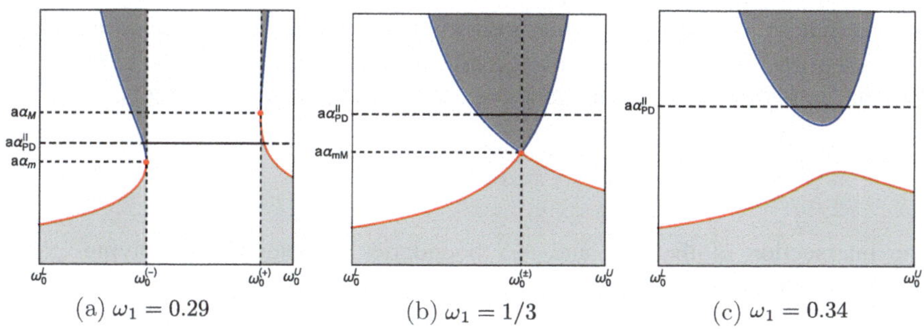

(a) $\omega_1 = 0.29$ (b) $\omega_1 = 1/3$ (c) $\omega_1 = 0.34$

Fig. 2.4 NS boundaries with three different values of ω_1

discriminant in the white region. The second condition is satisfied below the $a\alpha_{PD}^{II}$ boundary. The last condition is clearly satisfied if $a\alpha < 3/\omega_1$ where the PD boundary is below the $a\alpha = 3/\omega_1$ curve. In Fig. 2.4a, we selected $\omega_1 = 0.29 < 1/3$ with two distinct real roots, $\omega_0^{(-)} < \omega_0^{(+)}$. In this case, $1/4 < \omega_1 < 5/16$, so $a\alpha_m < a\alpha_{PD}^{II} < a\alpha_M$. Notice also that the red and blue curves are connected at points $(\omega_0^{(-)}, a\alpha_m)$ and $(\omega_0^{(+)}, a\alpha_M)$, which implies that the two curves form one NS boundary in interval $(\omega_0^L, \omega_0^{(-)}]$ and another one in interval $[\omega_0^{(+)}, \omega_0^U)$. If $\omega_0 \in (\omega_0^{(-)}, \omega_0^{(+)})$, then $D < 0$ and none of the NS boundaries are defined so stability is determined by the PD boundary $a\alpha_{PD}^{II}$. In Fig. 2.4b, the value of $\omega_1 = 1/3$ is chosen, when $D = 0$ and $\varphi(\omega_0, \omega_1, a\alpha)$ has one real root: $\omega_0^{(-)} = \omega_0^{(+)} = 5/9$ and therefore $a\alpha_m = a\alpha_M = 9/2 < a\alpha_{PD}^{II}$. The $a\alpha = a\alpha_{(+)}$ curve is colored in red, and it is convex and increasing before $\omega_0^{(-)} = \omega_0^{(+)}$ and convex and decreasing after $\omega_0^{(-)} = \omega_0^{(+)}$ with a kink at $\omega_0 = \omega_0^{(-)} = \omega_0^{(+)}$. The stable region is below the red NS boundary, which is colored in light gray. The curve $a\alpha = a\alpha_{(-)}$ is shown in blue color, convex and decreasing before $\omega_0 = \omega_0^{(-)} = \omega_0^{(+)}$ and convex and increasing as $\omega_0 > \omega_0^{(-)} = \omega_0^{(+)}$, with a kink at $\omega_0 = \omega_0^{(-)} = \omega_0^{(+)}$. The PD and NS boundaries cross at two points. The stable region is clearly above the (blue) NS boundary and below the PD boundary. The two stable regions have one common point at $\omega_0 = \omega_0^{(-)} = \omega_0^{(+)}$. In Fig. 2.4c, we selected $\omega_1 = 0.34 > 1/3$, when $D > 0$, $A < 0$ and $a\alpha_{(-)} > a\alpha_{(+)}$. The red $a\alpha = a\alpha_{(+)}$ curve is mound-shaped, while the blue $a\alpha = a\alpha_{(-)}$ curve is U-shaped. The two gray regions are separated. The PD boundary divides the darker (upper) region into two parts. As $\omega_1 > 3/8$, the $a\alpha_{PD}^{II}$ line has to be replaced by the $a\alpha = 3/\omega_1$ line. In summary of the above discussion, we have the following results.

Proposition 2.6 *Assume that* $\frac{1}{2}(1 - \omega_1) < \omega_0 < 1 - \omega_1$ *with* $0 < \omega_1 < 1$. *Then, the stability region is given as*

$$S = \left\{ (\omega_0, a\alpha) \mid \omega_0^L < \omega_0 < \omega_0^U,\ a\alpha < \min\left[\frac{2}{1 - 2\omega_1}, \frac{3}{\omega_1}\right] \text{ and } \varphi(\omega_0, \omega_1, a\alpha) > 0 \right\}.$$

The actual shape of this region depends on the choice of the value of ω_1.

We will now illustrate Propositions 2.5 and 2.6 with numerical examples. We first assume that $D < 0$, then $A < 0$ as well implying that $\varphi(\omega_0, \omega_1, a\alpha)$ is a convex parabola in $a\alpha$, and therefore the third condition of (2.22) holds for all $a\alpha$. If $D = 0$, then this condition holds for all values of $a\alpha$ except at the identical roots,

$$a\alpha_{(+)} = a\alpha_{(-)} = \frac{\omega_0 - 1}{2A}.$$

The intersection of the PD and NS boundaries is obtained by solving $\varphi(\omega_0, \omega_1, a\alpha) = 0$ with $a\alpha = a\alpha_{PD}^{II} = 2/(1 - 2\omega_1)$ to get

$$\omega_0^a = \frac{1}{2} \text{ and } \omega_0^b = \frac{5}{4} - 2\omega_1.$$

It is easy to show that for $\omega_1 < 1/3$,

$$\omega_0^a < \omega_0^{(-)} \text{ and } \omega_0^{(+)} < \omega_0^b$$

which follow from inequalities

$$\omega_0^{(-)} - \omega_0^a = \frac{1}{18}\left(2\sqrt{1-3\omega_1} - 1\right)^2 > 0$$

and

$$\omega_0^b - \omega_0^{(+)} = \frac{1}{36}\left(4\sqrt{1-3\omega_1} - 1\right)^2 > 0.$$

In addition, it can be shown that if $\omega_1 < 1/4$, then $\omega_0^b > \omega_0^U$ which is reversed for $\omega_1 > 1/4$.

In the following discussion, we will consider four sub-cases:

(A) $0 < \omega_1 \leq \frac{1}{4}$, when $\omega_0^U \leq \omega_0^b$.

(B) $\frac{1}{4} < \omega_1 \leq \frac{1}{3}$, when $\omega_0^U > \omega_0^b$.

(C) $\frac{1}{3} < \omega_1 \leq \frac{1}{2}$, when $a\alpha_{PD}^{II}$ is defined.

(D) $\frac{1}{2} < \omega_1 \leq 1$, when $a\alpha_{PD}^{II}$ is not defined.

Case (A): $0 < \omega_1 \leq \frac{1}{4}$

We have already proved that $a\alpha_{PD}^{II} < a\alpha_m < a\alpha_M$ in Fig. 2.3 implying that the PD boundary crosses the $a\alpha = a\alpha_{(+)}$ curve at points ω_0^a and ω_0^b. The domain of ω_0 is the interval $(0, 1-\omega_1)$ since $\omega_0^U < \omega_0^b$, and the stability region is bounded by the NS and PD boundaries which are given as

$$a\alpha = a\alpha_{(+)} \text{ if } 0 < \omega_0 < \omega_0^a$$

and

$$a\alpha = a\alpha_{PD}^{II} \text{ if } \omega_o^a < \omega_0 < 1 - \omega_1.$$

Figure 2.5a illustrates this case with $\omega_1 = 1/5$. The continuous black kinked curve is the partition curve. The blue points are the intersections of the NS and PD boundaries. The monopoly equilibrium loses stability through a PD or NS bifurcation depending on whether it crosses the PD or NS boundary. The two-parameter bifurcation diagram is shown in Fig. 2.5b with the same color code as in Fig. 2.1. Notice that the case of $\omega_0 < \omega_0^L$ was summarized in Proposition 2.5 and the other case of $\omega_0 > \omega_0^L$ in Proposition 2.6.

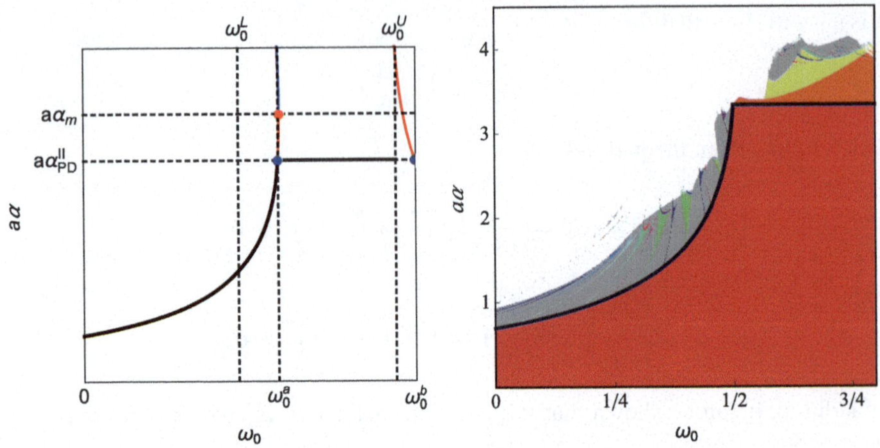

(a) Stability boundary

(b) Bifurcation diagram

Fig. 2.5 Dynamics for $0 < \omega_1 \leq 1/4$

Proposition 2.7 *If* $0 \leq \omega_1 < \dfrac{1}{4}$, *then the stability region is given as*

$$S_I^A = \left\{(\omega_0, a\alpha) \mid 0 < \omega_0 < 1 - \omega_1,\ a\alpha < \min\left[a\alpha_{(+)}, a\alpha_{PD}^{II}\right]\right\} \text{ if } D > 0$$

and

$$S_{II}^A = \left\{(\omega_0, a\alpha) \mid 0 < \omega_0 < 1 - \omega_1,\ a\alpha < a\alpha_{PD}^{II}\right\} \text{ if } D \leq 0.$$

Case (B): $\dfrac{1}{4} < \omega_1 \leq \dfrac{1}{3}$

In this case, we know that $\omega_0^U > \omega_0^b$ and from Fig. 2.3,

$$a\alpha_m < a\alpha_{PD}^{II} \leq a\alpha_M \text{ if } \frac{1}{4} < \omega_1 \leq \frac{5}{16}$$

and

$$a\alpha_m < a\alpha_M < a\alpha_{PD}^{II} \text{ if } \frac{5}{16} < \omega_1 \leq \frac{1}{3}.$$

Figure 2.6a illustrates the case with the choice of $\omega_1 = 0.32 > 5/16 = 0.3125$. The thick curve is the partition curve with three parts having kinks at the blue points. The values $\omega_0^{(-)}$ and $\omega_0^{(+)}$ are not labeled on the horizontal axis, and they are the abscissas of the red points. The union of the $a\alpha_{(+)}$ and $a\alpha_{(-)}$ curves is denoted by C_L for $\omega_0 \leq \omega_0^{(-)}$ and by C_R for $\omega_0 \geq \omega_0^{(+)}$. The third condition of (2.22) holds if the point $(\omega_0, a\alpha)$ is between these curves. Figure 2.6b shows a two-parameter bifurcation diagram.

$$a\alpha_{PD}^{II} \lesseqgtr \frac{3}{\omega_1} \text{ as } \omega_1 \lesseqgtr \frac{3}{8}.$$

If $\omega_1 < 3/8$ and $(\omega_0, a\alpha) \in S_{II}^C$, then $0 < a\alpha < a\alpha_{PD}^{II}$, $\varphi(\omega_0, \omega_1, a\alpha) > 0$ and $a\alpha < 3/\omega_1$ showing that all conditions of (2.22) are satisfied, so S_{II}^C is also a stability region. We will next show that S_{II}^C remains a stability region for $\omega_1 > 3/8$. Notice first that

$$a\alpha_{(-)} - \frac{3}{\omega_1} = -\frac{(1-\omega_0)\omega_1 + \omega_1\sqrt{D} + 6A}{2A\omega_1}. \tag{2.34}$$

The case of $A > 0$ has been examined in Proposition 2.5, so we assume that $A < 0$. The denominator of (2.34) is negative, and the numerator can be factored as

$$(1-\omega_0)\omega_1 + \omega_1\sqrt{D} - \frac{3}{2}\left((1-\omega_0)^2 - D\right) = \frac{3}{2}\left(\sqrt{D} + 1 - \omega_0\right)\left(\sqrt{D} - \left(1 - \omega_0 - \frac{2}{3}\omega_1\right)\right).$$

The first factor is positive. The second factor can be further factored:

$$\left[\omega_0 - \frac{1}{12}\left(9 - 8\omega_1 - \sqrt{3}\sqrt{3 - 8\omega_1}\right)\right]\left[\omega_0 - \frac{1}{12}\left(9 - 8\omega_1 + \sqrt{3}\sqrt{3 - 8\omega_1}\right)\right]$$

which is positive for $\omega_1 > 3/8$. Therefore,

$$a\alpha_{(-)} > \frac{3}{\omega_1} \text{ as } \omega_1 > \frac{3}{8},$$

implying that the third condition of (2.22) holds for $a\alpha > a\alpha_{(-)}$; however, the last condition is violated since

$$a\alpha > a\alpha_{(-)} > \frac{3}{\omega_1}.$$

The last condition holds for $a\alpha < 3/\omega_1$, but the third condition does not. Consequently, the stability conditions (2.22) cannot hold simultaneously if $\omega_1 > 3/8$. That is, S_{II}^C shrinks with increasing values of ω_1 and it disappears at $\omega_1 = 3/8$. Figure 2.7a shows the two disjoint stability regions with $\omega_1 = 0.34 < 3/8 = 0.375$, and Fig. 2.7b presents a two-parameter bifurcation diagram.

Proposition 2.9 *Assume that* $\frac{1}{3} < \omega_1 < \frac{1}{2}$, *and then the stability region of the monopoly equilibrium is*

$$S_I^C \cup S_{II}^C \textit{ if } \omega_1 < \frac{3}{8},$$

when S_I^C *and* S_{II}^C *are disjoint, and only*

$$S_I^C \textit{ if } \omega_1 > \frac{3}{8}.$$

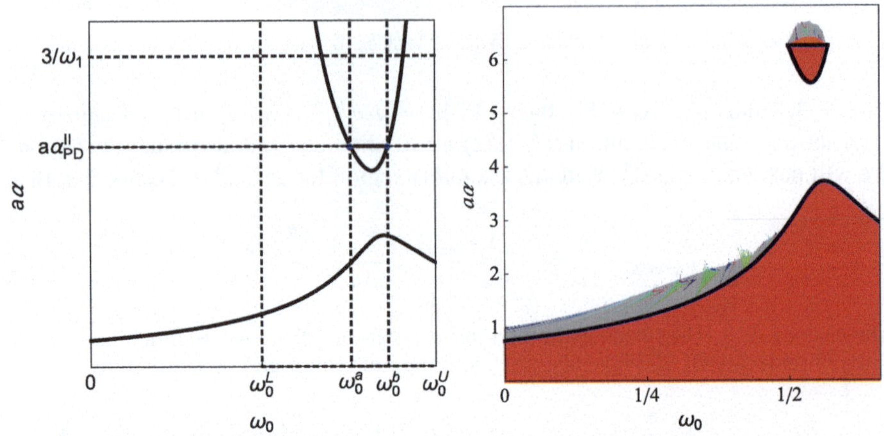

(a) Stability boundary (b) Bifurcation diagram

Fig. 2.7 Dynamics for $1/3 < \omega_1 \leq 1/2$

Case (D): $\dfrac{1}{2} < \omega_1 \leq 1$
A similar approach as shown before can verify the following result.

Proposition 2.10 *Assume that* $\dfrac{1}{2} < \omega_1 \leq 1$, *then the stability region of the monopoly equilibrium is*

$$S^D = \left\{(\omega_0, a\alpha) \mid 0 \leq \omega_0 \leq \omega_0^U,\ a\alpha < a\alpha_{(+)}\right\}.$$

Figure 2.8 illustrates a two-parameter bifurcation diagram for this case.
Finally, we examine the case of *geometric delay*, when

$$q^e(t+1) = \sum_{\tau=0}^{\infty} \omega(1-\omega)^{\tau} q(t-\tau). \tag{2.35}$$

Notice first that the sum of the coefficients is unity. Multiply the corresponding equation of $q^e(t)$ by $(1-\omega)$, and subtract the resulted equation from (2.35) to have

$$q^e(t+1) - (1-\omega)q^e(t) = \omega q(t).$$

Based on Eq. (2.3), we have the following two-dimensional difference equations:

$$\begin{aligned} q(t+1) &= q(t) + \alpha q(t)\left\{a - 2(b+c)\left[\omega q(t) + (1-\omega)q^e(t)\right]\right\}, \\ q^e(t+1) &= \omega q(t) + (1-\omega)q^e(t). \end{aligned} \tag{2.36}$$

The Jacobian of this system at the steady state $q = q^e = \bar{q}$ has the form

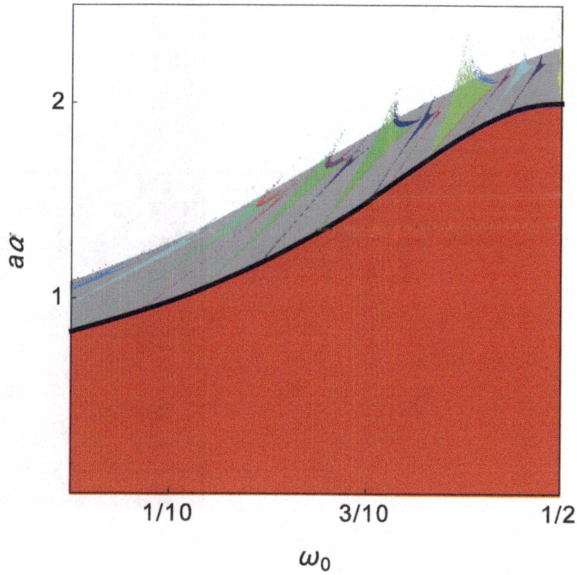

Fig. 2.8 Bifurcation diagram with $1/2 < \omega_1 \leq 1$

$$J = \begin{pmatrix} 1 - a\alpha\omega & -a\alpha(1-\omega) \\ \omega & 1-\omega \end{pmatrix}$$

with characteristic equation

$$\varphi(\lambda) = \lambda^2 + a_1\lambda + a_2 = 0$$

where

$$a_1 = -[2 - (1 + a\alpha)\omega] \text{ and } a_2 = 1 - \omega.$$

The conditions for local asymptotical stability are known to be (see Appendix C for details)

$$a_2 < 1$$

and

$$\pm a_1 + a_2 + 1 > 0.$$

In our case,

$$a_2 = 1 - \omega < 1$$

$$a_1 + a_2 + 1 = a\alpha\omega > 0$$

and

$$-a_1 + a_2 + 1 = 4 - (2 + a\alpha)\omega.$$

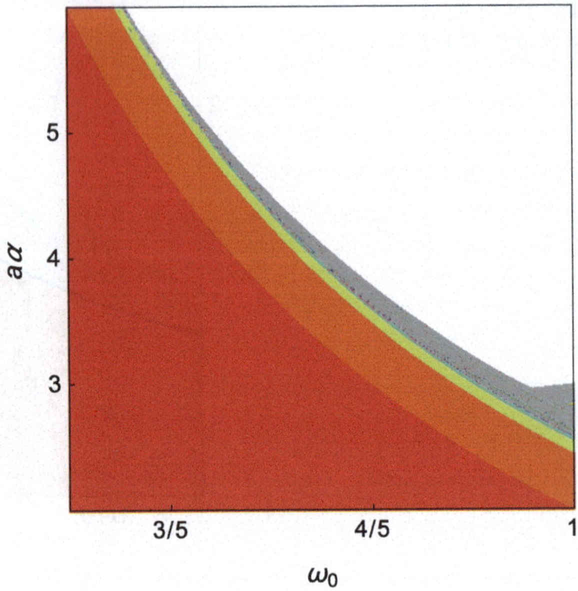

Fig. 2.9 Bifurcation diagram with geometric delay

The last condition holds if and only if

$$a\alpha < \frac{4-2\omega}{\omega}.$$

At this boundary value,

$$a_1 = 2-\omega \text{ and } \alpha_2 = 1-\omega$$

implying that

$$\varphi(\lambda) = (\lambda+1)\,(\lambda+(1-\omega))$$

showing that the eigenvalue $-(1-\omega)$ is inside the unit circle while the other eigenvalue is -1. Therefore at this boundary value, stability is lost through a PD bifurcation. Figure 2.9 shows a two-parameter bifurcation diagram for this case.

Proposition 2.11 *The stability region in the case of geometric delay is given as*

$$S_{III} = \left\{(\omega, a\alpha) \mid 0 < \omega \leq 1,\ a\alpha < a\alpha_{PD}^{III}\right\}$$

with $a\alpha_{PD}^{III}$ being the period doubling boundary:

$$a\alpha_{PD}^{III} = \frac{4-2\omega}{\omega}.$$

In the previous discussions, we always assumed that the firm is quantity adjusting. That is, its decision variable is the produced quantity and the unit price is its decreasing function. We can also consider monopolies when they select the unit prices for their products which then determine the market demands. The produced amount is the market demand to avoid overproduction or shortage. Mathematically, this is almost the same model as the one introduced and examined earlier in this section. Let p be the unit price of the product, then the market demand $d(p)$ is a decreasing function of p. So, the produced amount is $d(p)$, and therefore the revenue becomes $pd(p)$. The production cost is $C\left(d(p)\right)$, so the profit of the firm can be given as

$$\bar{\pi}(p) = pd(p) - C\left(d(p)\right)$$

which has the same form as (2.1): p instead of q, $d(p)$ replaces $p(q)$ and $C\left(d(p)\right)$ is instead of $C(q)$. Therefore, all methods and result discussed earlier can be applied without major differences.

2.2 Continuous Time Models with Fixed Delays

For the sake of simplicity, linear price and linear cost functions are considered:

$$p(q) = a - bq$$

and

$$C(q) = cq.$$

The profit of the firm equals

$$\pi(q) = q\left(a - bq\right) - cq \tag{2.37}$$

with marginal profit

$$\frac{\partial \pi(q)}{\partial q} = a - c - 2bq.$$

In order to avoid negative profit, we assume that $a > c$. As in the previous section, we assume that the firm adjusts its quantity in such a way that the relative output growth rate is proportional to the expected (or assessed) marginal profit:

$$\frac{\dot{q}(t)}{q(t)} = \alpha\left[a - c - 2bq^e(t)\right] \tag{2.38}$$

where $\alpha > 0$ is a speed of adjustment.

In the special case of complete information,

$$q^e(t) = q(t)$$

so (2.38) becomes

$$\dot{q}(t) = \alpha q(t)\,[a - c - 2bq(t)]\,. \tag{2.39}$$

This system has a meaningless zero steady state, and the unique positive steady state is

$$\bar{q} = \frac{a - c}{2b}.$$

Linearizing equation (2.39), we have

$$\frac{\partial \dot{q}(t)}{\partial q(t)} = \alpha\,(a - c - 2b\bar{q}) + \alpha\bar{q}(-2b) = -\alpha(a - c) < 0$$

implying that the monopoly equilibrium is always locally asymptotically stable.

Assume next that the firm's expected demand equals the realized demand at time $t - \tau$ $(\tau > 0)$, and then Eq. (2.39) is modified as

$$\dot{q}(t) = \alpha q(t)\,[a - c - 2bq(t - \tau)] \tag{2.40}$$

which has the same positive steady state $\bar{q}$ as the model without delay. Since at the steady state

$$\frac{\partial \dot{q}(t)}{\partial q(t)} = \alpha\,(a - c - 2b\bar{q}) = 0$$

and

$$\frac{\partial \dot{q}(t)}{\partial q(t - \tau)} = -\alpha\,(a - c) < 0,$$

with $q_\varepsilon = q - \bar{q}$, the linearized equation becomes

$$\dot{q}_\varepsilon(t) = -\alpha(a - c)q_\varepsilon(t - \tau). \tag{2.41}$$

Looking for the characteristic equation, we assume an exponential solution

$$q_\varepsilon(t) = e^{\lambda t}u$$

which is substituted into the equation,

$$\lambda e^{\lambda t}u = -\alpha(a - c)e^{\lambda(t-\tau)}u.$$

After simplification, we have

$$\lambda + \alpha(a - c)e^{-\lambda\tau} = 0 \tag{2.42}$$

which is a special case of Eq. (A2) of Appendix A with $P_0(\lambda) = \lambda$ and $P_1(\lambda) = \alpha(a-c)$. Notice first that all conditions $(i)-(iv)$ are satisfied. The system is locally asymptotically stable with $\tau = 0$, so stability is preserved for small values of τ. However, if the value of τ increases, then stability might be lost. At this critical value, at least one eigenvalue must have zero real part: $\lambda = iv$, where we can assume that $v > 0$, since complex conjugate of an eigenvalue is also an eigenvalue.

Substituting the λ value into (2.42), we have

$$iv + \alpha(a-c)\,(\cos v\tau - i \sin v\tau) = 0$$

and separating the real and imaginary parts,

$$\alpha(a-c)\cos v\tau = 0,$$

$$v - \alpha(a-c)\sin v\tau = 0,$$

so $\cos v\tau = 0$ and $\sin v\tau = 1$. Therefore,

$$v = \alpha(a-c)$$

and there are infinitely many solutions for τ:

$$\tau_n = \frac{1}{\alpha(a-c)}\left(\frac{\pi}{2} + 2n\pi\right) \text{ for } n = 0, 1, 2, \ldots \tag{2.43}$$

We will next determine the directions of stability switches at these values by selecting τ as the bifurcation parameter and considering the eigenvalue as function of τ, $\lambda = \lambda(\tau)$. By implicitly differentiating equation (2.42) with respect to τ, we get an equation for $\lambda'(\tau)$ as follows:

$$\lambda'(\tau) - \alpha(a-c)e^{-\lambda(\tau)\tau}\left(\lambda'(\tau)\tau + \lambda(\tau)\right) = 0$$

implying that

$$\lambda'(\tau) = \frac{\alpha(a-c)\lambda(\tau)e^{-\lambda(\tau)\tau}}{1-\alpha(a-c)\tau e^{-\lambda(\tau)\tau}}.$$

However from (2.42), $\alpha(a-c)e^{-\lambda(\tau)\tau} = -\lambda(\tau)$, so

$$\lambda'(\tau) = \frac{-\lambda(\tau)^2}{1+\tau\lambda(\tau)}.$$

At the critical value $\lambda = iv$, when

$$\lambda'(\tau) = \frac{v^2}{1+iv\tau} = \frac{v^2\,(1-iv\tau)}{1+(v\tau)^2}$$

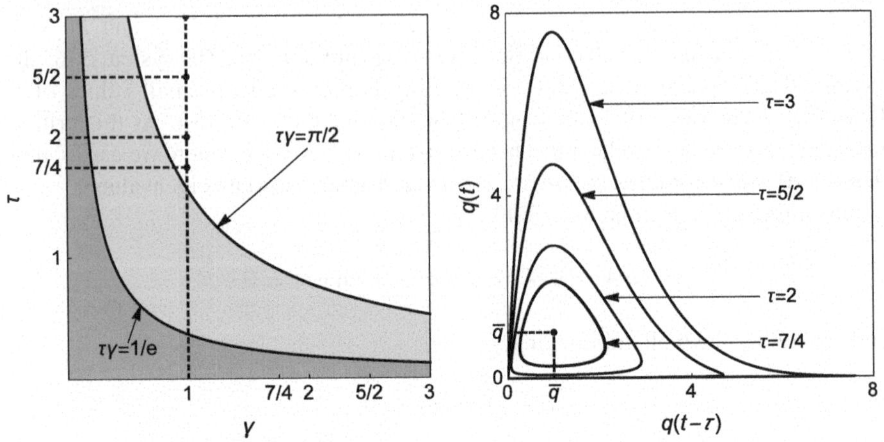

(a) Division of the (γ, τ) plane

(b) Birth of limit cycle

Fig. 2.10 Dynamics with one delay

with positive real part. So at the critical values of τ, at least one eigenvalue changes real part from negative to positive. Therefore, stability is lost at $\tau = \tau_0$ and stability cannot be regained later.

Proposition 2.12 *The monopoly equilibrium is locally asymptotically stable without delay, the stability is retained if* $\tau < \tau_0 = \frac{\pi}{2\alpha(a-c)}$, *and then stability is lost for all* $\tau > \tau_0$. *At* $\tau = \tau_0$, *Hopf bifurcation occurs giving the possibility of the birth of limit cycles.*

This result is illustrated in Fig. 2.10 where for the sake of simplicity we used notation $\gamma = \alpha(a - c)$.

In Figure 2.10a, we selected four points on the vertical line $\gamma = 1$ with $\tau = 7/4,\ 2,\ 5/2$ and 3, and in Fig. 2.10b the corresponding dynamics are shown. Clearly, the amplitude increases if the delay becomes larger. The downward sloping curve is the locus of $\tau\gamma = \pi/2$ which is the stability switching curve. We also show the $\tau\gamma = 1/e$ locus which is the partition curve between real and complex roots. The monopoly equilibrium is locally asymptotically stable in the gray region and unstable in the white region.

Assume next that $q^e(t)$ is formed as a weighted average of two past delays

$$q^e(t) = \omega q(t - \tau_1) + (1 - \omega)q(t - \tau_2) \tag{2.44}$$

where τ_1 and τ_2 are positive delays. Substituting this expression into (2.38), we have the delay differential equation,

$$\dot{q}(t) = \alpha q(t)\left[a - c - 2b\left(\omega q(t - \tau_1) + (1 - \omega)q(t - \tau_2)\right)\right]. \tag{2.45}$$

For examining local stability of the monopoly equilibrium, we first linearize this nonlinear equation. Notice that at the monopoly equilibrium

$$\frac{\partial \dot{q}(t)}{\partial q(t)} = \alpha \left[a - c - 2b\left(\omega \bar{q} + (1-\omega)\bar{q}\right)\right] = 0,$$

$$\frac{\partial \dot{q}(t)}{\partial q(t-\tau_1)} = -\alpha(a-c)\omega,$$

$$\frac{\partial \dot{q}(t)}{\partial q(t-\tau_2)} = -\alpha(a-c)\left(1-\omega\right).$$

Therefore, the linearized equation has the form

$$\dot{q}_\varepsilon(t) = -\alpha(a-c)\omega q_\varepsilon(t-\tau_1) - \alpha(a-c)\left(1-\omega\right) q_\varepsilon(t-\tau_2).$$

The characteristic polynomial can be obtained by substituting an exponential solution $q_\varepsilon(t) = e^{\lambda t}u$ into this equation,

$$\lambda e^{\lambda t} u = -\alpha(a-c)\omega e^{\lambda(t-\tau_1)} u - \alpha(a-c)\left(1-\omega\right) e^{\lambda(t-\tau_2)} u.$$

After simplification,

$$\lambda + \alpha(a-c)\omega e^{-\lambda\tau_1} + \alpha(a-c)\left(1-\omega\right) e^{-\lambda\tau_2} = 0. \tag{2.46}$$

This equation is a special case of (A6) of Appendix A with

$$P_0(\lambda) = \lambda,\ \ P_1(\lambda) = \alpha(a-c)\omega \text{ and } P_2(\lambda) = \alpha(a-c)(1-\omega).$$

Notice first that all conditions (i)–(iv) are satisfied. We will now use the method outlined in Appendix A to find the stability switching curve.

In this case, (2.46) can be rewritten as

$$1 + a_1(\lambda)e^{-\lambda\tau_1} + a_2(\lambda)e^{-\lambda\tau_2} = 0$$

with

$$a_1(\lambda) = \frac{\alpha(a-c)\omega}{\lambda} \text{ and } a_2(\lambda) = \frac{\alpha(a-c)(1-\omega)}{\lambda}.$$

Simple calculation shows that at $\lambda = iv$,

$$|a_1(iv)| = \frac{\alpha(a-c)\omega}{v} \text{ and } |a_2(iv)| = \frac{\alpha(a-c)(1-\omega)}{v},$$

so the feasible region of v is determined by relations (A11)–(A12) which have now the form

$$\frac{\alpha(a-c)\omega}{v} + \frac{\alpha(a-c)(1-\omega)}{v} \geq 1$$

and

$$-1 \leq \frac{\alpha(a-c)\omega}{v} - \frac{\alpha(a-c)(1-\omega)}{v} \leq 1$$

giving the relations

$$\frac{\alpha(a-c)}{v} \geq 1$$

and

$$-1 \leq \frac{\alpha(a-c)(2\omega-1)}{v} \leq 1$$

showing that

$$|\alpha(a-c)(2\omega-1)| \leq v \leq \alpha(a-c). \tag{2.47}$$

The case of $\omega = 1$ has been examined earlier in Eq. (2.41), so we can assume that $\omega < 1$. We can assume that $\omega \geq 1/2$; otherwise, we can interchange τ_1 and τ_2. First, we consider the case when $\omega > 1/2$. Then, (2.47) is an interval in the positive real line. From (A13) and (A14),

$$\begin{aligned}\theta_1 &= \cos^{-1}\left[\frac{1+\left(\frac{\alpha(a-c)\omega}{v}\right)^2 - \left(\frac{\alpha(a-c)(1-\omega)}{v}\right)^2}{2\frac{\alpha(a-c)\omega}{v}}\right], \\ &= \cos^{-1}\left[\frac{v^2+\alpha^2(a-c)^2(-1+2\omega)}{2\alpha(a-c)\omega v}\right]\end{aligned} \tag{2.48}$$

and

$$\begin{aligned}\theta_2 &= \cos^{-1}\left[\frac{1+\left(\frac{\alpha(a-c)(1-\omega)}{v}\right)^2 - \left(\frac{\alpha(a-c)\omega}{v}\right)^2}{2\frac{\alpha(a-c)(1-\omega)}{v}}\right], \\ &= \cos^{-1}\left[\frac{v^2+\alpha^2(a-c)^2(1-2\omega)}{2\alpha(a-c)(1-\omega)v}\right].\end{aligned} \tag{2.49}$$

Notice also that

$$\arg[a_1(iv)] = \arg[a_2(iv)] = \frac{3\pi}{2}$$

so from (A15) and (A16) we have the partition curves

$$\tau_1^{n\pm}(v) = \frac{1}{v}\left(\frac{3\pi}{2} + (2n-1)\pi \pm \theta_1\right) \tag{2.50}$$

Fig. 2.11 Stability switching curve in the (τ_1, τ_2) plane with n = 0

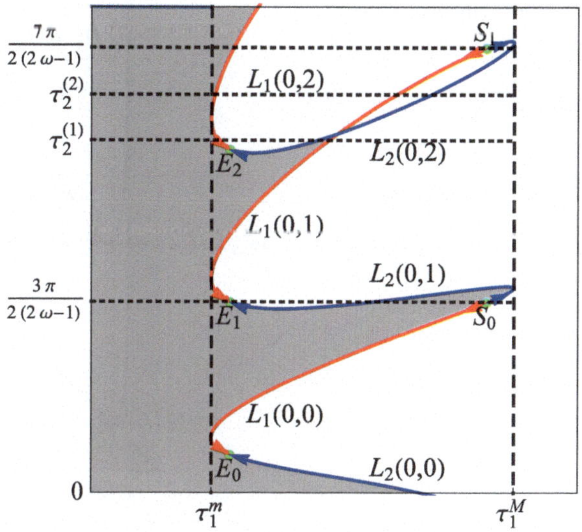

and

$$\tau_2^{m\mp}(v) = \frac{1}{v}\left(\frac{3\pi}{2} + (2m-1)\pi \mp \theta_2\right). \tag{2.51}$$

The continuous curves $(\tau_1^{n+}, \tau_2^{m-})$ and $(\tau_1^{n-}, \tau_2^{m+})$ in the (τ_1, τ_2) plane can be denoted as $L_1(n, m)$ and $L_2(n, m)$, respectively. They are illustrated in Fig. 2.11 with the selection of $\gamma = \alpha(a - c) = 1$.

At $\tau_1 = \tau_2 = 0$, the monopoly equilibrium is locally asymptotically stable. The segments of the stability switching curves with $n = 0$ are located in interval (τ_1^m, τ_1^M) where $\tau_1^m \simeq 1.493$ and $\tau_1^M \simeq 2.733$ with the parametric selection given before. The region where $0 \leq \tau_1 < \tau_1^m$ is part of the stability region, since the vertical line $\tau_1 = \tau_1^m$ has no intersection with the stability switching curve. The entire stability region is colored in gray.

Proposition 2.13 *If $\tau_1 < \tau_1^m$, then with any $\tau_2 > 0$, the delays are harmless. The curves $L_1(0, n)$ and $L_2(0, n)$ form a continuous stability switching curve. The stability region is the dark region in Fig. 2.11, and the white portion shows the instability region.*

Assume next that $\omega = 1/2$. Then, interval (2.47) is reduced to the following

$$0 < v \leq \alpha(a - c) \tag{2.52}$$

furthermore, from (2.48) and (2.49), we have

$$\theta_1 = \theta_2 = \cos^{-1}\left(\frac{v}{\alpha(a-c)}\right). \tag{2.53}$$

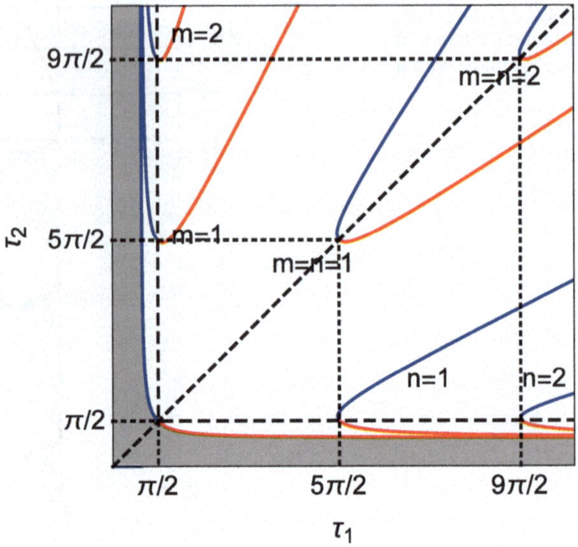

Fig. 2.12 Division of the (τ_1, τ_2) plane with $\omega = 1/2$

The corresponding stability switching curves $(\tau_1^{n+}, \tau_2^{m-})$ and $(\tau_1^{n-}, \tau_2^{m+})$ are obtained from Eqs. (2.50) and (2.51). The loci of $L_1(n, m)$ and $L_2(n, m)$ for $n = 0, 1$ and $m = 0, 1$ are illustrated in Fig. 2.12 with $\gamma = 1$.

In the special case of $m = n = 0$, the two loci form a hyperbola, which goes through point $(\pi/2, \pi/2)$. The stability region is the gray area under the hyperbola. Notice that the hyperbola is symmetric with respect to the 45° line, since the choice of $\omega = 1/2$ makes no difference between the two delays.

Proposition 2.14 *If $\tau_1 \leq 1$ or $\tau_2 \leq 1$, then regardless of the value of the other delay, the delays are harmless. The loci of $L_1(0, 0)$ and $L_2(0, 0)$ form a continuous curve, which is the stability switching curve. The stability region is the gray area under the hyperbola. The white area shows the instability region.*

In the model we examined before is called "gradient adjustment," since the relative growth rate of the output is proportional to the marginal profit. The expected (or assessed) output of the firm was either the current or delayed output or the convex linear combination of two past output values. Finally, we examine the special case when instead of the second delayed output value the current output is used:

$$q^e(t) = \omega q(t - \tau) + (1 - \omega)q(t) \ (0 < \omega < 1) \tag{2.54}$$

which is called *adaptive expectation*. So, Eq. (2.38) has the following form:

$$\dot{q}(t) = \alpha q(t) \left[a - c - 2b \left(\omega q(t - \tau) + (1 - \omega)q(t)\right)\right]. \tag{2.55}$$

It is a special case of system (2.45) with $\tau_1 = \tau$ and $\tau_2 = 0$. Instead of specializing the two-delay solution, we can derive the stability condition directly. Notice first that at the monopoly equilibrium $\bar{q}$,

$$\frac{\partial \dot{q}(t)}{\partial q(t)} = -\alpha(a-c)(1-\omega)$$

and

$$\frac{\partial \dot{q}(t)}{\partial q(t-\tau)} = -\alpha(a-c)\omega.$$

So, the linearized equation is the following:

$$\dot{q}_\varepsilon(t) + \alpha(a-c)(1-\omega)q_\varepsilon(t) + \alpha(a-c)\omega q_\varepsilon(t-\tau) = 0. \tag{2.56}$$

Substituting the exponential solution $q_\varepsilon(t) = e^{\lambda t}u$ into this equation, we have

$$\lambda e^{\lambda t} + \alpha(a-c)(1-\omega)e^{\lambda t}u + \alpha(a-c)\omega e^{\lambda(t-\tau)}u = 0$$

and after simplifications, the characteristic equation becomes

$$\lambda + \alpha(a-c)(1-\omega) + \alpha(a-c)\omega e^{-\lambda\tau} = 0. \tag{2.57}$$

At $\tau = 0$, we have $\lambda = -\alpha(a-c) < 0$, so the system is stable. If τ increases, then stability might be lost when $\lambda = iv$ $(v > 0)$, which is then substituted into this equation

$$iv + \alpha(a-c)(1-\omega) + \alpha(a-c)\omega\,(\cos v\tau - i\sin v\tau) = 0.$$

Separation of the real and imaginary parts shows that

$$\alpha(a-c)\omega \sin v\tau = v \tag{2.58}$$

and

$$\alpha(a-c)\omega\cos v\tau = -\alpha(a-c)(1-\omega) \tag{2.59}$$

showing that $\sin v\tau > 0$ and $\cos v\tau < 0$. Adding the squares of these equations shows that

$$\alpha^2(a-c)^2\omega^2 = v^2 + \alpha^2(a-c)^2(1-\omega)^2$$

so

$$v^2 = \alpha^2(a-c)^2(2\omega - 1).$$

If $\omega \leq 1/2$, then no positive solution exists for v^2; otherwise,

$$v = \alpha(a-c)\sqrt{2\omega - 1}. \tag{2.60}$$

From (2.58), we have the critical values,

$$\bar{\tau}_n = \frac{1}{\alpha(a-c)\sqrt{2\omega-1}}\left[\pi - \sin^{-1}\left(\frac{\sqrt{2\omega-1}}{\omega}\right) + 2n\pi\right].$$

The direction of the stability switch can be determined by Hopf bifurcation. As before, the eigenvalue is considered as a function of the bifurcation parameter τ, $\lambda = \lambda(\tau)$. By implicitly differentiating the characteristic equation with respect to τ, we have

$$\lambda' + \alpha(a-c)\omega e^{-\lambda\tau}(-\lambda'\tau - \lambda) = 0$$

implying that

$$\lambda' = \frac{\alpha(a-c)\omega\lambda e^{-\lambda\tau}}{1 - \alpha(a-c)\omega\tau e^{-\lambda\tau}}.$$

From (2.57), we know that

$$\alpha(a-c)\omega e^{-\lambda\tau} = -\lambda - \alpha(a-c)(1-\omega),$$

so

$$\lambda' = \frac{-\lambda^2 - \alpha(a-c)(1-\omega)\lambda}{1 + \lambda\tau + \alpha(a-c)(1-\omega)\tau}.$$

At $\lambda = iv$, this relation becomes

$$\lambda' = \frac{v^2 - \alpha(a-c)(1-\omega)iv}{1 + \alpha(a-c)(1-\omega)\tau + iv\tau}\,\frac{1 + \alpha(a-c)(1-\omega)\tau - iv\tau}{1 + \alpha(a-c)(1-\omega)\tau - iv\tau}.$$

Since the denominator becomes positive, Re λ' has the same sign as

$$v^2\left[1 + \alpha(a-c)(1-\omega)\tau\right] - \alpha(a-c)(1-\omega)\tau v^2 = v^2 > 0$$

showing that at each critical value $\bar{\tau}_n$, at least one pair of eigenvalues changes the sign of the real part from negative to positive.

Proposition 2.15 *The monopoly equilibrium with adaptive expectation is locally asymptotically stable if* $\omega \le \frac{1}{2}$. *Otherwise, it is locally stable if* $\tau < \bar{\tau}_0$, *locally unstable if* $\tau > \bar{\tau}_0$ *and at the critical value* $\tau = \bar{\tau}_0$, *Hopf bifurcation is observed implying that the stability loss occurs with the birth of limit cycles.*

Notice that

$$\frac{\sqrt{2\omega-1}}{\omega} = \sqrt{\frac{2\omega-1}{\omega^2}}$$

which is an increasing function, since

$$\left(\frac{2\omega - 1}{\omega^2}\right)' = \frac{2\omega(1-\omega)}{\omega^4} > 0,$$

and as ω runs through interval $(1/2, 1)$, its value changes from 0 to 1. In this interval, $\sin^{-1}$ is also increasing, so $\bar{\tau}_0$ strictly decreases in both ω and α. So, the increase of any (or both) of these values has a decreasing effect on $\bar{\tau}_0$, so the stability interval $[0, \bar{\tau}_0]$ becomes shorter making the monopoly equilibrium less stable.

2.3 Continuous Time Models with Distributed Delays

We turn our attention to the same monopoly model as the one discussed in the previous section; however instead of fixed delay, continuously distributed delay is assumed. This is the case where the delay is uncertain, so it can be considered as a random variable, or the firm wants to react to an average of past information instead of sudden market change. Therefore, it is assumed that in the expected marginal profit (2.38),

$$q^e(t) = \int_0^t W(t-s, m, T) q(s) ds \tag{2.61}$$

where the gamma density-type weighting function is

$$W(t-s, m, T) = \begin{cases} \frac{1}{T} e^{-\frac{t-s}{T}} & \text{if } m = 0, \\ \frac{1}{m!}\left(\frac{m}{T}\right)^{m+1} (t-s)^m e^{-\frac{m(t-s)}{T}} & \text{if } m \geq 1. \end{cases} \tag{2.62}$$

So, Eq. (2.39) becomes the following:

$$\dot{q}(t) = \alpha q(t) \left[a - c - 2b \int_0^t W(t-s, m, T) q(s) ds\right]. \tag{2.63}$$

After linearization, we have

$$\dot{q}_\varepsilon(t) = -2\alpha b \bar{q} q_\varepsilon^e(t) \tag{2.64}$$

where

$$q_\varepsilon^e(t) = \int_0^t W(t-s, m, T) q_\varepsilon(s) ds. \tag{2.65}$$

As usual, we are looking for the solution as the exponential form $q_\varepsilon(t) = e^{\lambda t} u$, and after substituting it into Eq. (2.64), we get

$$\lambda + 2\alpha b\bar{q} \int_0^t W(t-s, m, T) e^{-\lambda(t-s)} ds = 0.$$

Introduce the new integration variable, $z = t - s$ as it is shown in Appendix B, and then

$$\int_0^t W(t-s, m, T) e^{-\lambda(t-s)} ds = \int_0^t W(z, m, T) e^{-\lambda z} dz.$$

As $t \to \infty$, the right hand side converges to

$$\int_0^\infty W(z, m, T) e^{-\lambda z} dz = \begin{cases} (1 + \lambda T)^{-1} & \text{if } m = 0, \\ \left(1 + \frac{\lambda T}{m}\right)^{-(m+1)} & \text{if } m \geq 1, \end{cases}$$

as it is proved in Appendix B. For the sake of simplicity, introduce variables $\gamma = 2\alpha b\bar{q} = \alpha(a - c) > 0$ and

$$\bar{m} = \begin{cases} 1 & \text{if } m = 0, \\ m & \text{if } m \geq 1 \end{cases}$$

then the characteristic polynomial becomes

$$\lambda + \gamma \left(1 + \frac{\lambda T}{\bar{m}}\right)^{-(m+1)} = 0$$

which can be rewritten as

$$\gamma + \lambda \left(1 + \frac{\lambda T}{\bar{m}}\right)^{m+1} = 0. \tag{2.66}$$

This is an $(m + 2)$nd degree polynomial of λ. The monopoly equilibrium is locally asymptotically stable if all eigenvalues have negative real parts. Theorem B.1 of Appendix B presents sufficient and necessary conditions for this based on the Hurwitz matrices. For larger values of m, it is very difficult to check the stability conditions in general. Therefore, we will consider smaller values of m and examine stability of the monopoly equilibrium.

Assume first that $m = 0$ and $T > 0$. Then Eq. (2.66) becomes

$$\gamma + \lambda(1 + \lambda T) = 0$$

which is a quadratic equation in λ,

$$T\lambda^2 + \lambda + \gamma = 0. \tag{2.67}$$

Since all coefficients are positive, both eigenvalues have negative real parts, so the monopoly equilibrium is locally asymptotically stable with all $T > 0$. In the case of fixed delays, the stability condition was the requirement that the delay had to be sufficiently small,

$$\tau < \frac{\pi}{2\gamma}.$$

We can therefore say that system with fixed delay is less stable.

Assume next that $m = 1$ and $T > 0$. Then, Eq. (2.66) has the form

$$\gamma + \lambda(1 + T\lambda)^2 = T^2\lambda^3 + 2T\lambda^2 + \lambda + \gamma = 0 \tag{2.68}$$

which is a cubic equation in λ:

$$a_0\lambda^3 + a_1\lambda^2 + a_2\lambda + a_3 = 0$$

with

$$a_0 = T^2,\ a_1 = 2T,\ a_2 = 1 \text{ and } a_3 = \gamma.$$

Since all coefficients are positive, Theorem B.1 implies that the roots of (2.68) are negative or have negative real parts if and only if

$$a_1a_2 > a_3a_0.$$

In our case,

$$a_1a_2 - a_3a_0 = T(2 - \gamma T) > 0$$

if and only if

$$T < \frac{2}{\gamma} = T_1^*. \tag{2.69}$$

The $T = 2/\gamma$ hyperbola divides the (γ, T) plane into two parts, the stability region is under the hyperbola given in gray color, and the white region above the hyperbola is the instability region as shown in Fig. 2.13.

We will next show that at the critical value of T_1^*, Hopf bifurcation occurs confirming the birth of limit cycle. Consider T as the bifurcation parameter and the eigenvalue as the function of T, $\lambda = \lambda(T)$. Implicit differentiation of equation (2.68) with respect to T shows that

$$\lambda' + 2\lambda^2 + 4T\lambda\lambda' + 2T\lambda^3 + 3T^2\lambda^2\lambda' = 0$$

implying that

$$\lambda' = -\frac{2\lambda^2 + 2T\lambda^3}{1 + 4T\lambda + 3T^2\lambda^2}. \tag{2.70}$$

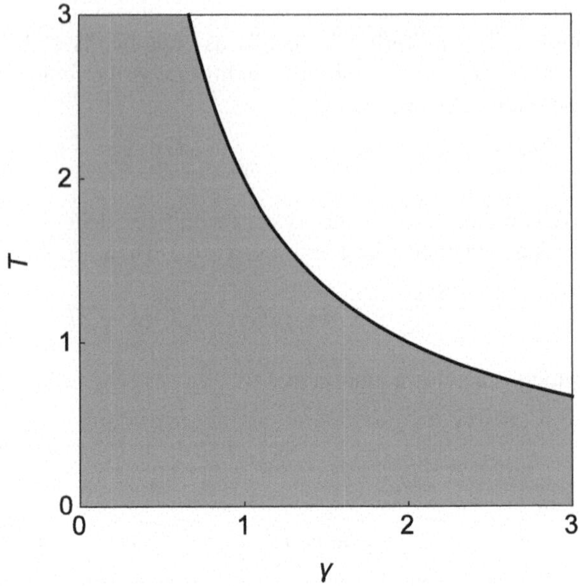

Fig. 2.13 Stability region in the case of m = 1

At the critical T value, the characteristic polynomial simplifies as

$$0 = T^2\lambda^3 + 2T\lambda^2 + \lambda + \frac{2}{T} = \left(\lambda + \frac{2}{T}\right)\left(1 + \lambda^2 T^2\right).$$

So, the roots are

$$\lambda_1 = -\frac{2}{T} \text{ and } \lambda_{2,3} = \pm i\frac{1}{T}.$$

At $\lambda = \lambda_2$ (case of $\lambda = \lambda_3$ is similar),

$$\lambda' = \frac{2+2i}{T^2(-2+4i)}\frac{-2-4i}{-2-4i}$$

with positive real part,

$$\text{Re}\left(\lambda'\right) = \frac{1}{5T^2} > 0. \tag{2.71}$$

Hence, the real part of the eigenvalue changes from negative to positive resulting in stability loss on the partition curve and the occurrence of Hopf bifurcation.

We turn our attention to the case of $m = 2$ and $T > 0$. The characteristic equation

$$\gamma + \lambda\left(1 + \frac{\lambda T}{2}\right)^3 = \gamma + \lambda + \frac{3}{2}\lambda^2 T + \frac{3}{4}\lambda^3 T^2 + \frac{1}{8}\lambda^4 T^3 = 0 \tag{2.72}$$

is equivalent with a quartic equation

$$a_0\lambda^4 + a_1\lambda^3 + a_2\lambda^2 + a_3\lambda + a_4 = 0$$

with

$$a_0 = T^3,\ a_1 = 6T^2,\ a_2 = 12T,\ a_3 = 8,\ a_4 = 8\gamma.$$

The Hurwitz matrix has now the form from (B7):

$$\boldsymbol{H}_4 = \begin{pmatrix} \frac{6}{T} & 1 & 0 & 0 \\ \frac{8}{T^3} & \frac{12}{T^2} & \frac{6}{T} & 1 \\ 0 & \frac{8\gamma}{T^3} & \frac{8}{T^3} & \frac{12}{T^2} \\ 0 & 0 & 0 & \frac{8\gamma}{T^3} \end{pmatrix}$$

with principal minors

$$\boldsymbol{H}_1 = \frac{6}{T},\ \boldsymbol{H}_2 = \frac{64}{T^3},\ \boldsymbol{H}_3 = \frac{32\,(16 - 9T\gamma)}{T^6} \text{ and } \boldsymbol{H}_4 = \frac{8\gamma}{T^3}\boldsymbol{H}_3.$$

Therefore, all eigenvalues have negative real parts if and only if

$$T < \frac{16}{9\gamma} = T_2^*. \tag{2.73}$$

Similarly to the previous case, it is easy to show that at $T = T_2^*$ stability is lost and Hopf bifurcation occurs.

In the case of $m = 3$ and $T > 0$, the characteristic polynomial is

$$\gamma + \lambda\left(1 + \frac{\lambda T}{3}\right)^4 = \gamma + \lambda + \frac{4}{3}\lambda^2 T + \frac{6}{9}\lambda^3 T^2 + \frac{4}{27}\lambda^4 T^3 + \frac{1}{81}\lambda^5 T^4 = 0 \tag{2.74}$$

which is equivalent with a fifth degree equation

$$a_0\lambda^5 + a_1\lambda^4 + a_2\lambda^3 + a_3\lambda^2 + a_4\lambda + a_5 = 0$$

with

$$a_0 = T^4,\ a_1 = 12T^3,\ a_2 = 54T^2,\ a_3 = 108T,\ a_4 = 81,\ a_5 = 81\gamma.$$

Simple calculation shows that the first three principal minors of the Hurwitz matrix are positive, and the fourth and fifth have the same sign as $-\gamma^2T^2 - 336\gamma T + 576$. This is a concave parabola with roots

$$T_{1,2} = \frac{24\left(-7 \pm 5\sqrt{2}\right)}{\gamma}.$$

One root is positive, and the other is negative. The value of the parabola is positive if

$$T < \frac{24\left(-7 + 5\sqrt{2}\right)}{\gamma} = T_3^*. \tag{2.75}$$

We can also show that at this critical value T_3^*, stability is lost via Hopf bifurcation.

The case of $m \to \infty$ is discussed next. Equation (2.66) can be rewritten as

$$\gamma + \lambda\left(1 + \frac{T\lambda}{m}\right)^m \left(1 + \frac{T\lambda}{m}\right) = 0$$

and as $m \to \infty$, it becomes

$$\gamma + \lambda e^{T\lambda} = 0 \text{ or } \lambda + \gamma e^{-T\lambda} = 0.$$

Notice that this equation is identical to (2.42), which is the characteristic equation with fixed delay. In that case, the critical value was

$$T_0 = \frac{\pi}{2\gamma}.$$

Clearly,

$$T_0 \approx \frac{1.5708}{\gamma} < T_3^* \approx \frac{1.7056}{\gamma} < T_2^* \approx \frac{1.7778}{\gamma} < T_1^* = \frac{2}{\gamma}.$$

For $m = 1, 2, 3$ and ∞, the partition curves are shown in Fig. 2.14 showing that fixed delay has the least stable system and as m decreases, the monopoly equilibrium becomes more stable since the stability region increases. We can summarize the previously shown results as follows.

Proposition 2.16 *The monopoly equilibrium is locally asymptotically stable with* $m = 0$ *for all* $T > 0$. *As* $m = 1, 2, 3$ *and as* $m \to \infty$, *the stability switching curves are hyperbolic in* γ. *The stability region shrinks as m increases, and the* $m \to \infty$ *case (fixed delay) has the least stable equilibrium.*

In the previous discussion, it was assumed that the firm selects a weighted average of its past output values as its expectation formation. A different type of expectation is given by the *adaptive rule*,

$$q^e(t) = \omega q^e(t) + (1 - \omega)q(t)$$

where

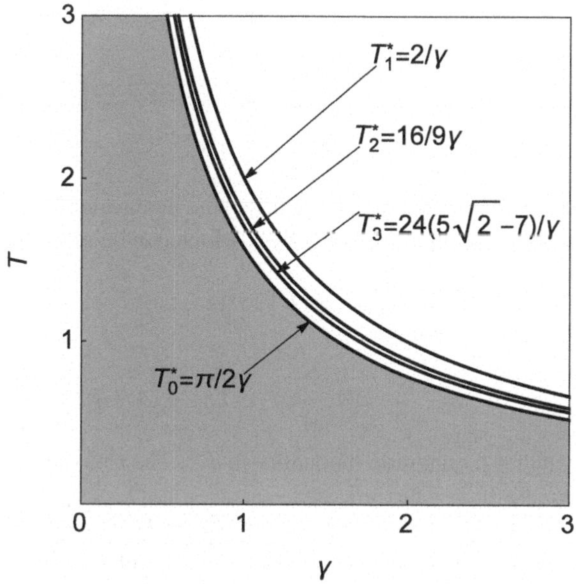

Fig. 2.14 Three partition curves and stability region

$$q^e(t) = \int_0^t W(t-s, T, m)q(s)ds. \tag{2.76}$$

This is a two-stage process. First, the weighted average of all past output values is determined to get $q^e(t)$, and then the expected demand is chosen between this weighted average and the current output level.

Similarly to the characteristic equation (2.66), it is easy to show that in this case it is modified as follows:

$$\gamma\omega + (\lambda + \gamma(1-\omega))\left(1 + \frac{\lambda T}{\bar{m}}\right)^{m+1} = 0. \tag{2.77}$$

Instead of dealing with this general higher degree polynomial, special cases of small values of m are examined.

In the case of $m = 0$ and $T > 0$, the characteristic equation becomes quadratic,

$$\lambda^2 T + (1 + T\gamma(1-\omega))\,\lambda + \gamma = 0.$$

Since all coefficients are positive, the monopoly equilibrium is locally asymptotically stable for all $T > 0$.

Assume next that $m = 1$ and $T > 0$. Then (2.77) becomes

$$\gamma\omega + (\lambda + \gamma(1-\omega))\left(1 + 2\lambda T + \lambda^2 T^2\right) = 0$$

which is a cubic equation

$$a_0\lambda^3 + a_1\lambda^2 + a_2\lambda + a_3 = 0$$

with

$$a_0 = T^2,\ a_1 = 2T + \gamma(1-\omega)T^2,\ a_2 = 1 + 2T\gamma(1-\omega) \text{ and } a_3 = \gamma.$$

Since all coefficients are positive, the eigenvalues are negative or have negative real parts if and only if $a_1a_2 > a_0a_3$ which can be rewritten as

$$a_1a_2 - a_0a_3 = T\left[2\gamma^2(1-\omega)^2T^2 + (4\gamma - 5\gamma\omega)\,T + 2\right] > 0$$

or

$$2(1-\omega)^2\,(\gamma T)^2 + (4-5\omega)\,(\gamma T) + 2 > 0 \tag{2.78}$$

which is a quadratic inequality in $T\gamma$. The roots of the left hand side are

$$\gamma T_A = \frac{5\omega - 4 - \sqrt{\omega(9\omega - 8)}}{4(1-\omega)^2}$$

and

$$\gamma T_B = \frac{5\omega - 4 + \sqrt{\omega(9\omega - 8)}}{4(1-\omega)^2}.$$

The discriminant is negative if $\omega < 8/9$, and then (2.78) holds for all $T\gamma > 0$ implying the local asymptotical stability of the monopoly equilibrium. If $\omega = 8/9$, then the left hand side of (2.78) becomes

$$\frac{2}{81}\,(\gamma T)^2 - \frac{4}{9}\,(\gamma T) + 2 = 2\left(\frac{\gamma T}{9} - 1\right)^2$$

which is positive if $\gamma T \neq 9$. Assume next that $\omega > 8/9$. Then, both roots are positive and (2.78) holds if either

$$0 < \gamma T < \gamma T_A \text{ or } \gamma T_B < \gamma T. \tag{2.79}$$

Figure 2.15 shows the stability region in the $(\omega, \gamma T)$ plane where the white area is the stability region and the instability region is colored in gray.

It is clear from the figure that the locus of $a_1a_2 - a_0a_3 = 0$ is the partition curve. The $\omega = \bar{\omega} > 8/9$ vertical line crosses the partition curve at points $A = (\bar{\omega}, \gamma T_A)$ and $B = (\bar{\omega}, \gamma T_B)$, and the monopoly equilibrium is locally asymptotically stable if either $0 < \gamma T < \gamma T_A$ or $\gamma T_B < \gamma T$ as the analytic derivations suggest. If $\omega = 8/9$, then the point $C = (8/9, 9)$ represents the identical roots of the left hand side of (2.78). In the special case of $\omega = 1$, inequality (2.78) is specified as

$$-\gamma T + 2 > 0$$

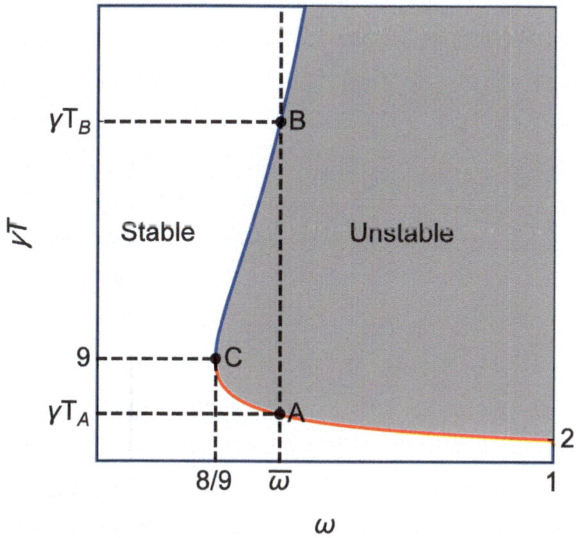

Fig. 2.15 Instability region in case of m = 1 is colored in gray

which can be rewritten as $\gamma T < 2$. Notice that this condition is the same as (2.69). Similarly to the earlier cases, it can be proved that

$$\text{Re}\,\lambda' > 0 \text{ at } \gamma T = \gamma T_A \text{ and } \text{Re}\,\lambda' < 0 \text{ at } \gamma T = \gamma T_B.$$

Proposition 2.17 *At the critical value γT_A, stability is lost and it is regained at $\gamma T = \gamma T_B$. In both cases, Hopf bifurcation occurs.*

Figure 2.16 shows the bifurcation diagram in the $(\gamma T, q(t))$ plane with $\omega = 0.91$. Assume next that $m = 2$. Then, Eq. (2.77) has the form

$$\gamma\omega + (\lambda + \gamma(1-\omega))\left(1 + \frac{\lambda T}{2}\right)^3$$
$$= \gamma\omega + (\lambda + \gamma(1-\omega))\left(1 + \frac{3}{2}\lambda T + \frac{3}{4}\lambda^2 T^2 + \frac{1}{8}\lambda^3 T^3\right) = 0$$

which is a quartic polynomial equation

$$a_0\lambda^4 + a_1\lambda^3 + a_2\lambda^2 + a_3\lambda + a_4 = 0$$

with

$$a_0 = \frac{T^3}{8},\ a_1 = \frac{3T^2}{4} + \gamma(1-\omega)\frac{T^3}{8},\ a_2 = \frac{3T}{2} + \gamma(1-\omega)\frac{3T^2}{4}$$

$$a_3 = 1 + \frac{3T\gamma(1-\omega)}{2},\ a_4 = \gamma.$$

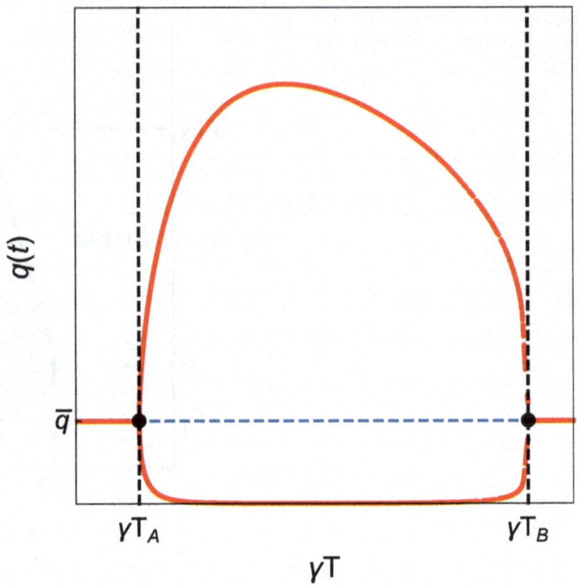

Fig. 2.16 Bifurcation diagram along the $\omega = 0.91$ line

The Hurwitz matrix has the form

$$\boldsymbol{H} = \begin{pmatrix} a_1 & a_0 & 0 & 0 \\ a_3 & a_2 & a_1 & a_0 \\ 0 & a_4 & a_3 & a_2 \\ 0 & 0 & 0 & a_4 \end{pmatrix}.$$

Notice first that

$$\boldsymbol{H}_1 = a_1 = \frac{3T^2}{4} + \gamma(1-\omega)\frac{T^3}{8} > 0,$$

$$\boldsymbol{H}_2 = a_1 a_2 - a_0 a_3$$
$$= \left(\tfrac{3T^2}{4} + \gamma(1-\omega)\tfrac{T^3}{8}\right)\left(\tfrac{3T}{2} + \gamma(1-\omega)\tfrac{3T^2}{4}\right) - \tfrac{T^3\gamma}{8}\left(1 + \tfrac{3T\gamma(1-\omega)}{2}\right) > 0,$$

$$\boldsymbol{H}_4 = a_4 \boldsymbol{H}_3$$

and since $a_4 > 0$, we need to deal only with the sign of $\boldsymbol{H}_3$. Simple calculation shows that

$$\boldsymbol{H}_3 = \alpha_0(\gamma T)^3 + \alpha_1(\gamma T)^2 + \alpha_2(\gamma T) + \alpha_3 \tag{2.80}$$

with

$$\alpha_0 = (9\omega - 8)(1-\omega)^2,\ \alpha_1 = -12(4 - 9\omega + 5\omega^2),\ \alpha_2 = -12(8 - 11\omega),\ \alpha_3 = -64.$$

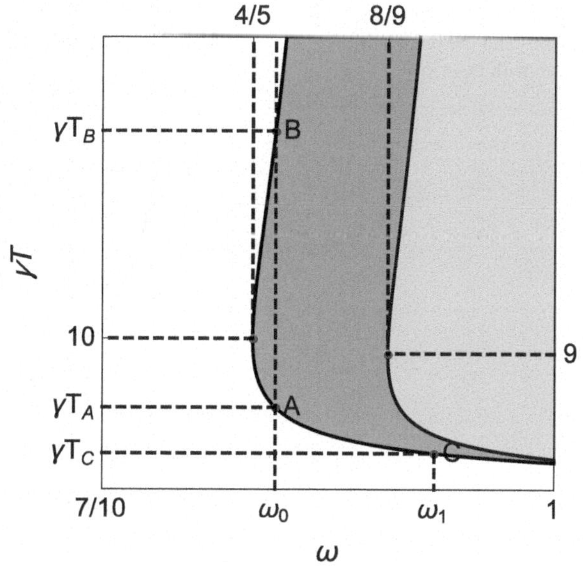

Fig. 2.17 Partition curves with m = 1 and m = 2

The locus of $\boldsymbol{H}_3 = 0$ in the $(\omega, \gamma T)$ plane is shown in Fig. 2.17.

It divides the plane into stability and instability regions. The light gray area is the instability region for $m = 1$, and by adding the dark gray area to it gives the instability region for $m = 2$. The value of $\omega_0 = 0.815$ is selected. Equation $\boldsymbol{H}_3 = 0$ has two positive roots, $\gamma T_A \simeq 5.42$ (shown as point A in the figure) and $\gamma T_B \simeq 23.75$ (shown as point B), in addition to a negative root. Stability occurs if $0 < \gamma T < \gamma T_A$ or $\gamma T_B < \gamma T$, and the unstable interval is $(\gamma T_A, \gamma T_B)$. It can be proved that stability is lost at $\gamma T = \gamma T_A$ and regained at $\gamma T = \gamma T_B$. If $\omega = 8/9$, then the cubic polynomial of $\boldsymbol{H}_3$ is reduced to a quadratic polynomial, and if $8/9 < \omega < 1$, then the cubic equation $\boldsymbol{H}_3 = 0$ has only one positive root, which is denoted by point C on the figure. If γT is increased from zero along the vertical line $\omega = \omega_1$, then at point C stability is lost. In all cases, Hopf bifurcation occurs.

For $m > 2$, numerical simulation is used to find the stability switching curves. They are shown in Fig. 2.18.

In the figure, the $(n-1)$st principal minor of $\boldsymbol{H}_n$ is denoted by D_n^{n-1} for $n = 3, 4, 5, 6, 7$. The left most curve is the partition curve with $n \to \infty$ which is model (2.55) examined earlier. It was shown that the partition curve was

$$\gamma \bar{T}_0 = \frac{1}{\sqrt{2\omega - 1}} \left[\pi - \sin^{-1} \left(\frac{\sqrt{2\omega - 1}}{\omega} \right) \right].$$

As m increases, the stability region shrinks making the monopoly equilibrium less stable.

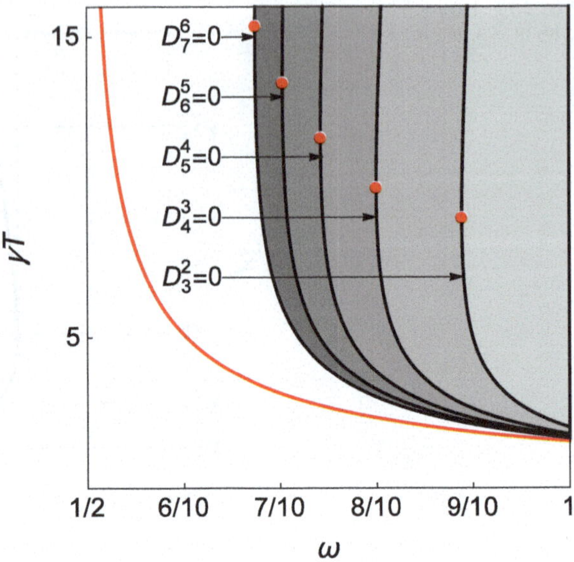

Fig. 2.18 Instability gray regions and the red partition curve with m→ ∞

In the previous models, we assumed that $q^e(t)$ is a weighted average output level from zero to t. With increasing value of t, this interval also increases, and for large values of t there is no reason to go back far behind in output values in computing averages. Therefore, it is more reasonable to introduce *bounded continuously distributed delays*, when

$$q^e(t) = \frac{1}{I}\int_{t-\Delta}^{t} \frac{1}{T}e^{-\frac{t-s}{T}}q(s)ds \tag{2.81}$$

where

$$I = \int_{t-\Delta}^{t} \frac{1}{T}e^{-\frac{t-s}{T}}ds = \int_{0}^{\Delta} \frac{1}{T}e^{-\frac{u}{T}}du = 1 - e^{-\frac{\Delta}{T}}$$

with the new integration variable $u = t - s$. So, the weighting function is

$$\frac{1}{IT}e^{-\frac{t-s}{T}}$$

which has unit integral value in interval $(t - \Delta, t)$. Introduce function

$$z(t) = \int_{t-\Delta}^{t} \frac{1}{T}e^{-\frac{t-s}{T}}q(s)ds. \tag{2.82}$$

For the sake of simplicity, we consider the simple version of Eq. (2.38) as

$$\dot{q}(t) = \alpha\left[a - c - 2bq^e(t)\right] = \alpha\left[a - c - \frac{2b}{I}z(t)\right] \tag{2.83}$$

and combining it with

$$\dot{z}(t) = \frac{1}{T}(q(t) - z(t)) - \frac{1}{T}e^{-\frac{\Delta}{T}}q(t - \Delta) \tag{2.84}$$

we obtained a delay two-dimensional system. Notice that from (2.83),

$$z(t) = \frac{I}{2\alpha b}[\alpha(a - c) - \dot{q}(t)]$$

and so

$$\dot{z}(t) = -\frac{I}{2\alpha b}\ddot{q}(t).$$

After we substitute this relation into (2.84), a second-order one-dimensional delay equation is obtained.

$$-\frac{I}{2\alpha b}\ddot{q}(t) = \frac{1}{T}\left[q(t) - \frac{I}{2\alpha b}[\alpha(a - c) - \dot{q}(t)] - e^{-\frac{\Delta}{T}}q(t - \Delta)\right]$$

which can be simplified as follows:

$$T\ddot{q}(t) + \dot{q}(t) + \frac{2\alpha b}{I}q(t) - \frac{2\alpha b}{I}e^{-\frac{\Delta}{T}}q(t - \Delta) - \alpha(a - c) = 0. \tag{2.85}$$

For the sake of simplicity, introduce the notation

$$A = \frac{2\alpha b}{I} \text{ and } B = \alpha(a - c).$$

We have

$$T\ddot{q}(t) + \dot{q}(t) + Aq(t) - Ae^{-\frac{\Delta}{T}}q(t - \Delta) - B = 0. \tag{2.86}$$

Notice first that the steady state of the system is again $\bar{q} = (a - c)/2b$. As before, in finding the characteristic equation, we look for the solution in an exponential form, $q(t) = e^{\lambda t}u$, and substitute it into the homogeneous version of (2.86),

$$T\lambda^2 e^{\lambda t}u + \lambda e^{\lambda t}u + Ae^{\lambda t}u - Ae^{-\frac{\Delta}{T}}e^{\lambda(t-\Delta)}u = 0$$

and after simplifications,

$$T\lambda^2 + \lambda + A - Ae^{-\frac{\Delta}{T}-\lambda\Delta} = 0. \tag{2.87}$$

Stability switch occurs when $\lambda = iv$ $(v > 0)$ and by substitution

$$-Tv^2 + iv + A - Ae^{-\frac{\Delta}{T}}(\cos\Delta v - i\sin\Delta v) = 0.$$

Separation of the real and imaginary parts gives the following relations

$$Ae^{-\frac{\Delta}{T}} \cos \Delta v = A - Tv^2, \tag{2.88}$$

$$Ae^{-\frac{\Delta}{T}} \sin \Delta v = -v. \tag{2.89}$$

By adding the squares of these equations,

$$A^2 e^{-\frac{2\Delta}{T}} = \left(A - Tv^2\right)^2 + v^2. \tag{2.90}$$

This relation required that

$$A^2 > \left(A - Tv^2\right)^2 + v^2$$

which can be rewritten as

$$v^2 < \frac{2AT - 1}{T^2}. \tag{2.91}$$

Solution exists if

$$AT > \frac{1}{2}. \tag{2.92}$$

Inequality (2.91) is quadratic in T,

$$T^2 v^2 - 2AT + 1 < 0$$

which holds if T is between the roots of the left hand side:

$$\frac{A - \sqrt{A^2 - v^2}}{v^2} < T < \frac{A + \sqrt{A^2 - v^2}}{v^2}. \tag{2.93}$$

The discriminant is positive by (2.90).

Next, we show that any solution $\lambda = iv$ is a single root. Otherwise, it would also satisfy equation

$$2T\lambda + 1 + A\Delta e^{-\frac{\Delta}{T} - \lambda\Delta} = 0 \tag{2.94}$$

in addition to (2.87). Multiplying (2.87) by Δ and adding the resulting equation to (2.94) to have

$$\Delta T \lambda^2 + (\Delta + 2T)\lambda + \Delta A + 1 = 0.$$

If $\lambda = iv$, then

$$\left(-\Delta T v^2 + \Delta A + 1\right) + (\Delta + 2T)\, iv = 0.$$

The imaginary part is zero at $v = 0$, in which case the first term is positive. This is a contradiction.

Equation (2.90) implies that

$$T^2 v^4 - 2ATv^2 + A^2\left(1 - e^{-\frac{2\Delta}{T}}\right) + v^2 = 0$$

which is a quadratic equation of T with discriminant

$$D = 4v^4\left(A^2 e^{-\frac{2\Delta}{T}} - v^2\right).$$

So, solution exists only for

$$v \leq Ae^{-\frac{\Delta}{T}}$$

and the solutions are

$$T_{1,2} = \frac{A \pm \sqrt{A^2 e^{-\frac{2\Delta}{T}} - v^2}}{v^2} \quad (T_1 \leq T_2)\,. \tag{2.95}$$

They are separated by the curve A/v^2.

Both roots are positive since $e^{-\frac{2\Delta}{T}} < 1$, and they satisfy relation (2.93). In the special case of $v = 0$, both roots converge to infinity, and as $v = Ae^{-\frac{\Delta}{T}}$, the discriminant is zero so $T_1 = T_2$. The roots $T_1(v)$ and $T_2(v)$ as functions of v are shown in Fig. 2.19, where we use $\beta = \Delta/T$ as fixed. Solution exists if $0 \leq v \leq Ae^{-\beta}$.

Notice that T_2 strictly decreases in v, and simple differentiation shows that T_1 is increasing if

$$v^4 + 4A^2v^2\left(1 - e^{-\frac{2\Delta}{T}}\right) - 4A^4 e^{-\frac{2\Delta}{T}}\left(1 - e^{-\frac{2\Delta}{T}}\right) > 0.$$

Fig. 2.19 Graphs of functions $T_1(\nu)$ and $T_2(\nu)$

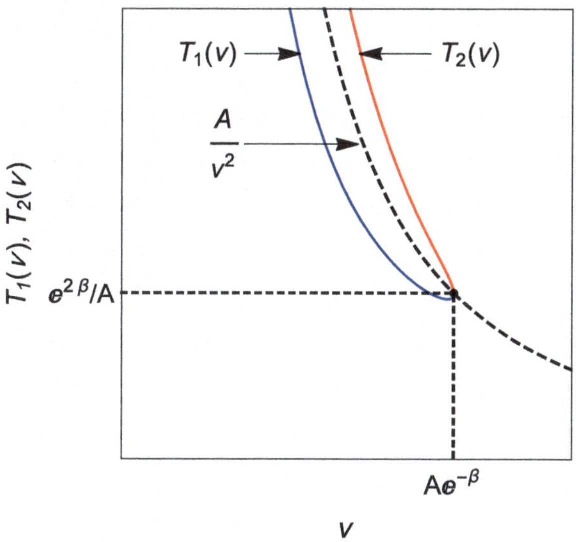

One root of the left hand side is negative, and the other is positive for v^2, so the only positive solution for v is

$$v^* = A\sqrt{2\left(\sqrt{1-e^{-\frac{2\Delta}{T}}} - \left(1-e^{-\frac{2\Delta}{T}}\right)\right)}. \tag{2.96}$$

So, T_1 increases if $v > v^*$ and decreases if $v < v^*$. It is also easy to see that $v^* < Ae^{-\frac{\Delta}{T}}$, so solutions in (2.95) exist with $v = v^*$.

From (2.89), we have

$$\sin(\beta T v) = -\frac{v}{A}e^{\beta}$$

therefore

$$\bar{T}_1^{(n)} = \frac{1}{\beta v}\left[-\sin^{-1}\left(\frac{v}{A}e^{\beta}\right) + 2n\pi\right] \quad (n \geq 1) \tag{2.97}$$

if $A - \bar{T}_1^{(n)} v^2 \geq 0$ or

$$\bar{T}_2^{(m)} = \frac{1}{\beta v}\left[\sin^{-1}\left(\frac{v}{A}e^{\beta}\right) + (2m+1)\pi\right] \quad (m \geq 0) \tag{2.98}$$

if $A - \bar{T}_2^{(m)} v^2 < 0$.

These two cases are due to Eq. (2.88) and sign of $\cos \Delta v$. Figure 2.20 shows the graphs of $\bar{T}_1^{(n)}$ and $\bar{T}_2^{(m)}$ for $n = 1, 2, 3, 4, 5$ and $m = 0, 1, 2, 3, 4$, when $\bar{T}_1^{(n)}$ are

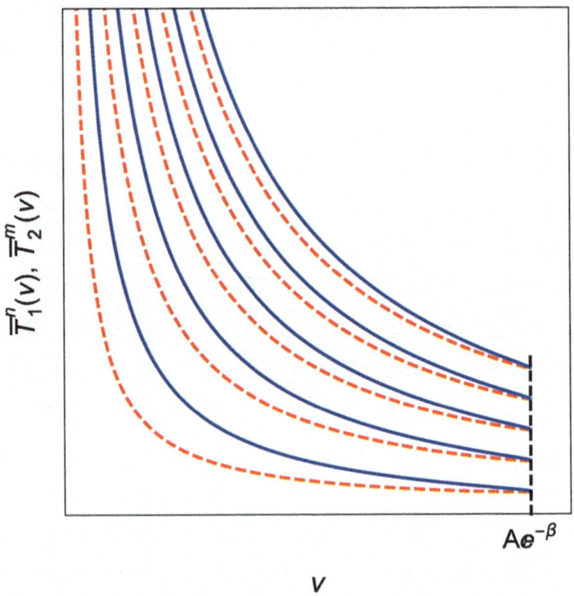

Fig. 2.20 Shapes of $\bar{T}_1^n(\nu)$ and $\bar{T}_2^m(\nu)$ for $1 \leq n \leq 5$ and $0 \leq m \leq 4$

shown with solid line and $\bar{T}_2^{(m)}$ with broken lines. It is also clear that $\bar{T}_2^{(k)}(v)$ and $\bar{T}_1^{(k+1)}(v)$ have the common end point at $v = Ae^{-\beta}$.

This is clear, since solutions $\bar{T}_1$ and $\bar{T}_2$ are equal at this value.

Proposition 2.18 *Stability switches may occur when curves $T_1(v)$ and $\bar{T}_1^{(n)}(v)$ or $T_2(v)$ and $\bar{T}_2^{(m)}(v)$ intersect with $n \geq 1$ and $m \geq 0$.*

Since

$$\sin^{-1}\left(\frac{v}{A}e^{\beta}\right) \in \left(0, \frac{\pi}{2}\right),$$

we also have

$$\bar{T}_2^{(0)}(v) < \bar{T}_1^{(1)}(v) < \bar{T}_2^{(1)}(v) < \bar{T}_1^{(2)}(v) < \bar{T}_2^{(2)}(v) < \bar{T}_1^{(3)}(v) < \ldots \text{for all } v \in \left(0, Ae^{-\beta}\right).$$

Figures 2.21, 2.22 and 2.23 show the intersections with $\beta = 1, 1/2$ and 2, respectively, where $A = 4$ is chosen.

The direction of stability switches can be examined by selecting T as the bifurcation parameter, considering the eigenvalue as function of $T : \lambda = \lambda(T)$. Then implicitly differentiating equation (2.87) with respect to T with keeping $\beta = \Delta/T$,

$$\lambda^2 + 2T\lambda\lambda' + \lambda' - Ae^{-\beta-\lambda\beta T}(-\lambda'\beta T - \lambda\beta) = 0$$

implying that

Fig. 2.21 Case of $\beta = 1$

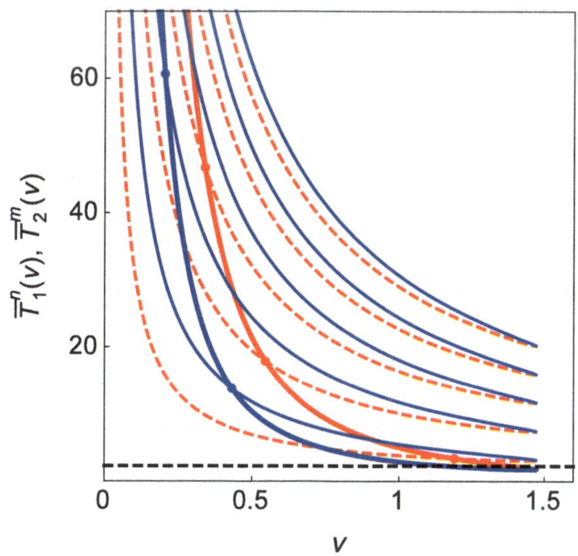

Fig. 2.22 Case of $\beta = 1/2$

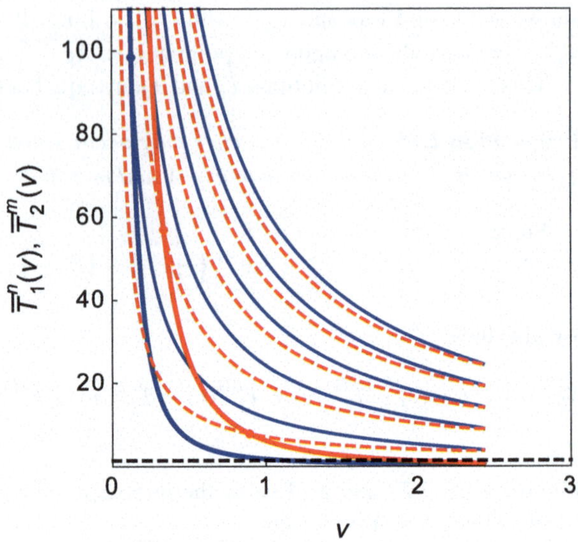

Fig. 2.23 Case of $\beta = 2$

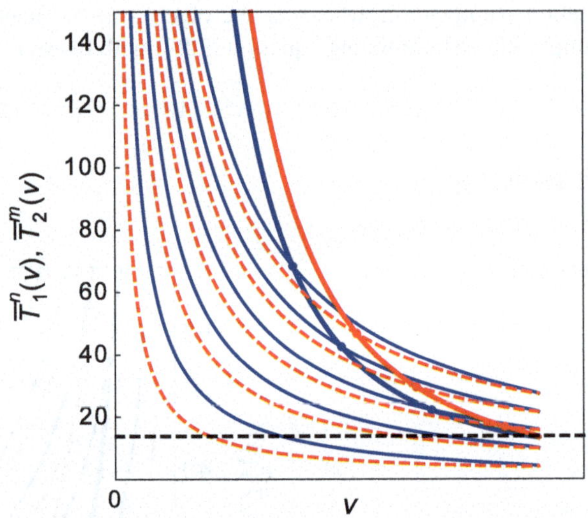

$$\lambda' = -\frac{\lambda^2 + A\beta\lambda e^{-\beta-\lambda\beta T}}{2T\lambda + 1 + A\beta T e^{-\beta-\lambda\beta T}}$$

$$= -\frac{\lambda^2 + \beta\lambda\left(T\lambda^2 + \lambda + A\right)}{2T\lambda + 1 + \beta T\left(T\lambda^2 + \lambda + A\right)}$$

where Eq. (2.87) was used. As $\lambda = iv$,

$$\lambda' = \frac{v^2(1+\beta) + i\left(\beta T v^3 - A\beta v\right)}{\left(-v^2 T^2 \beta + 1 + \beta A T\right) + iv\left(2T + \beta T\right)}.$$

A lengthy examination of the sign of Re λ' shows the following result.

Proposition 2.19 *There is a positive threshold*

$$T^* = \frac{A\beta^2 + \sqrt{A^2\beta^4 + 4(\beta+1)A^2\beta^2 e^{-2\beta}}}{2A^2\beta^2 e^{-2\beta}} > \frac{1}{A}e^{2\beta}$$

such that: (a) If $T < T^*$, then at any intersection of $T_1(v)$ and $\bar{T}_1^{(n)}(v)$ stability is lost; (b) if $T > T^*$, then at any intersection of $T_1(v)$ and $\bar{T}_1^{(n)}(v)$ stability might be regained; (c) at any intersection of $T_2(v)$ and $\bar{T}_2^{(m)}(v)$, stability is lost; (d) at any intersection, only one pair of eigenvalues changes the sign of its real part.

2.4 Supplementary Notes and Discussions

Very few works have been devoted to monopolies; however, all results developed to oligopolies can be applied in this special case. The earliest developments up to 1975 are summarized in Okuguchi (1976), and their multiproduct generalizations are presented in Okuguchi and Szidarovszky (1999). The discussed models are all linear, where local asymptotical stability implies global asymptotical stability. The recent developments on nonlinear oligopolies are presented in Bischi et al. (2010).

Section 2.1 is mainly based on Matsumoto and Szidarovszky (2014a) where a detailed analysis is given with discrete time scales. Section 2.2 is based on two papers of Matsumoto and Szidarovszky (2012a, 2014b), and the two-delay cases are developed based on the method of Gu et al. (2005), the details of which are given in Appendix A. Section 2.3 is a further developed version of the results presented in Matsumoto and Szidarovszky (2014b). Matsumoto et al. (2013) introduced bounded continuously distributed delays, which are discussed in the last part of this section. An elementary study of monopolies with continuous time scales and a single delay is discussed in Matsumoto and Szidarovszky (2012b). Multiple continuously distributed delays are described in Matsumoto and Szidarovszky (2015b), while some nonlinear extensions are given in Matsumoto and Szidarovszky (2014c, d).

Chapter 3
Dynamic Duopolies

Consider two firms which produce identical or related products or offer identical or related services to a homogeneous market. Their objective is to have as much profit as possible.

Let q_1 and q_2 denote the production outputs of the firms. Since the products are related, the unit price of each product depends on the supplies of both, so the unit prices depend on both q_1 and q_2:

$$p_1 = p_1(q_1, q_2) \text{ and } p_2 = p_2(q_1, q_2).$$

Let $C_1(q_1)$ and $C_2(q_2)$ denote the production cost functions, then the profit of firm k is given as the difference of its revenue and cost,

$$\pi_k(q_1, q_2) = q_k p_k(q_1, q_2) - C_k(q_k). \tag{3.1}$$

Similarly to the monopoly case, this model also might include other competitors with known total production levels which do not change in time. If L_k is the capacity limit of firm k, then the set of feasible output values of firm k is the closed interval $[0, L_k]$. In this model, the firms select output levels as their decision variables. Such competitions are called *Cournot duopolies*.

The firms might also select the unit prices p_1 and p_2 for their products, and then the market demands determine the produced output levels. In the mathematical description of the price-adjusting models, consider the following equations:

$$p_1 = p_1(q_1, q_2) \text{ and } p_2 = p_2(q_1, q_2) \tag{3.2}$$

and assume that for all positive pairs (p_1, p_2), there is a unique solution for q_1 and q_2,

$$q_1 = d_1(p_1, p_2) \text{ and } q_2 = d_2(p_1, p_2).$$

A. Matsumoto and F. Szidarovszky, *Dynamic Oligopolies with Time Delays*,
https://doi.org/10.1007/978-981-13-1786-6_3

Functions d_1 and d_2 are usually called the *market demand* functions, and the price functions are called the *inverse demand* functions. The profit of firm k can then be given as

$$\bar{\pi}_k(p_1, p_2) = d_k(p_1, p_2)p_k - C_k(d_k(p_1, p_2)). \tag{3.3}$$

These types of competitions are called *Bertrand duopolies.*

In most duopoly models discussed in the literature, linear cost functions are assumed, when

$$C_1(q_1) = c_1 q_1 \text{ and } C_2(q_2) = c_2 q_2.$$

As the price functions are concerned, there are two major classes of duopoly models.

In the case of *linear price functions*, it is assumed that

$$p_k = \alpha_k - \beta_k q_k - \gamma_k q_\ell \quad (k = 1, 2 \text{ and } \ell \neq k) \tag{3.4}$$

where $\alpha_k > 0$ is the maximum price, and coefficients β_k and γ_k measure the substitutability between the products. If $\gamma_k = 0$ for $k = 1, 2$, then the products are independent, and if $\alpha_1 = \alpha_2$ and $\beta_1 = \beta_2 = \gamma_1 = \gamma_2$, then the products are perfect substitutes, when the model is a duopoly without product differentiation. Otherwise, the model is a duopoly with product differentiation. Notice that Eq. (3.1) has the form

$$\pi_k(q_1, q_2) = q_k(\alpha_k - \beta_k q_k - \gamma_k q_\ell - c_k) \tag{3.5}$$

and the marginal profit equals

$$\frac{\partial \pi_k(q_1, q_2)}{\partial q_k} = \alpha_k - 2\beta_k q_k - \gamma_k q_\ell - c_k. \tag{3.6}$$

In order to find the profits of the Bertrand firms, we have to consider equation (3.2),

$$\begin{aligned} p_1 &= \alpha_1 - \beta_1 q_1 - \gamma_1 q_2 \\ p_2 &= \alpha_2 - \beta_2 q_2 - \gamma_2 q_1 \end{aligned} \tag{3.7}$$

which is a system of linear equations. It is easy to show that the solutions for q_1 and q_2 are as follows:

$$\begin{aligned} q_1 &= \frac{1}{\beta_1\beta_2 - \gamma_1\gamma_2}(\alpha_1\beta_2 - \alpha_2\gamma_1 - p_1\beta_2 + \gamma_1 p_2) = d_1(p_1, p_2) \\ q_2 &= \frac{1}{\beta_1\beta_2 - \gamma_1\gamma_2}(\alpha_2\beta_1 - \alpha_1\gamma_2 - p_2\beta_1 + \gamma_2 p_1) = d_2(p_1, p_2) \end{aligned} \tag{3.8}$$

where we have to assume that the denominator is nonzero. Therefore, the profit of firm k has the special form,

$$\bar{\pi}_k(p_1, p_2) = (p_k - c_k)\,\frac{1}{\beta_1\beta_2 - \gamma_1\gamma_2}\,(\alpha_k\beta_\ell - \alpha_\ell\gamma_k - p_k\beta_\ell + \gamma_k p_\ell) \tag{3.9}$$

where $k = 1, 2$ and $\ell \neq k$. The marginal profit of the firm is therefore

$$\frac{\partial\bar{\pi}_k(p_1, p_2)}{\partial p_k} = \frac{1}{\beta_1\beta_2 - \gamma_1\gamma_2}\,(\alpha_k\beta_\ell - \alpha_\ell\gamma_k - 2p_k\beta_\ell + \gamma_k p_\ell + c_k\beta_\ell) \tag{3.10}$$

which has the same form as (3.6) where $\alpha_k,\ \beta_k,\ \gamma_k$ and c_k are replaced by

$$\frac{\alpha_k\beta_\ell + c_k\beta_\ell}{\beta_1\beta_2 - \gamma_1\gamma_2},\ \frac{\beta_\ell}{\beta_1\beta_2 - \gamma_1\gamma_2},\ \frac{-\gamma_k}{\beta_1\beta_2 - \gamma_1\gamma_2}\ \text{and}\ \frac{\alpha_\ell\gamma_k}{\beta_1\beta_2 - \gamma_1\gamma_2},$$

respectively.

The other class of duopoly models are based on *hyperbolic* price functions:

$$p_1 = \frac{1}{q_1 + \theta_1 q_2}\ \text{and}\ p_2 = \frac{1}{q_2 + \theta_2 q_1}\quad (\theta_1, \theta_2 \in [0, 1]) \tag{3.11}$$

where θ_1 and θ_2 show the degree of product substitutability. If $\theta_k = 1$ for $k = 1, 2$, then the products are perfect substitutes, in which case there is no product differentiation. If $\theta_k = 0$ for $k = 1, 2$, then the two products are independent. To avoid trivial cases, we assume that $0 < \theta_1, \theta_2 < 1$. The profit of firm k is the following:

$$\pi_k = \frac{q_k}{q_k + \theta_k q_\ell} - c_k q_k\ (k = 1, 2,\ \ell \neq k) \tag{3.12}$$

with the marginal profit,

$$\frac{\partial\pi_k(q_1, q_2)}{\partial q_k} = \frac{\theta_k q_\ell}{(q_k + \theta_k q_\ell)^2} - c_k. \tag{3.13}$$

If the competition is a Bertrand duopoly, then we need to determine the market demand functions by solving equations (3.2), which now have the following forms:

$$p_1 = \frac{1}{q_1 + \theta_1 q_2}\ \text{and}\ p_2 = \frac{1}{q_2 + \theta_2 q_1}\quad (\theta_1, \theta_2 \in (0, 1)). \tag{3.14}$$

The solution is obtained as

$$q_1 = \frac{p_2 - \theta_1 p_1}{(1 - \theta_1\theta_2)p_1 p_2}\ \text{and}\ q_2 = \frac{p_1 - \theta_2 p_2}{(1 - \theta_1\theta_2)p_1 p_2} \tag{3.15}$$

where the denominator is positive.

The profit of the Bertrand firms is given as

$$\begin{aligned}\bar{\pi}_k(p_1, p_2) &= (p_k - c_k)\frac{p_\ell - \theta_k p_k}{(1-\theta_1\theta_2)p_1 p_2} \\ &= \frac{p_k p_\ell - \theta_k p_k^2 - c_k p_\ell + c_k\theta_k p_k}{(1-\theta_1\theta_2)p_1 p_2} \\ &= \frac{1}{1-\theta_1\theta_2}\left(1 - \theta_k\frac{p_k}{p_\ell} - \frac{c_k}{p_k} + \theta_k\frac{c_k}{p_\ell}\right) \quad (k = 1, 2,\ \ell \neq k)\end{aligned} \tag{3.16}$$

with marginal profits

$$\frac{\partial\bar{\pi}_k(p_1, p_2)}{\partial p_k} = \frac{1}{1-\theta_1\theta_2}\left(-\frac{\theta_k}{p_\ell} + \frac{c_k}{p_k^2}\right) \quad (k = 1, 2,\ \ell \neq k). \tag{3.17}$$

In the cases of Cournot duopolies, both firms select output levels as decision variables, while in the cases of Bertrand duopolies, both firms select prices as decision variables.

In the linear Cournot duopoly from (3.6), we have

$$\frac{\partial^2\pi_k(q_1, q_2)}{\partial q_k^2} = -2\beta_k < 0$$

and in the case of linear Bertrand duopoly from (3.10),

$$\frac{\partial^2\bar{\pi}_k(p_1, p_2)}{\partial p_k^2} = -\frac{2\beta_\ell}{\beta_1\beta_2 - \gamma_1\gamma_2}$$

which is negative if $\beta_1\beta_2 - \gamma_1\gamma_2 > 0$. This is a logical assumption since the demand of a product has larger effect on its price than the demand of the other product.

In the hyperbolic cases, we have from (3.13) that

$$\frac{\partial^2\pi_k(q_1, q_2)}{\partial q_k^2} = -\frac{2\theta_k q_\ell}{(q_k + \theta_k q_\ell)^3} < 0$$

and from (3.17),

$$\frac{\partial^2\bar{\pi}_k(p_1, p_2)}{\partial p_k^2} = -\frac{2c_k}{(1-\theta_1\theta_2)p_k^3} < 0$$

under our assumption that $0 < \theta_1, \theta_2 < 1$. Therefore, the profit function of each firm is strictly concave in its own decision variable in both Cournot and Bertrand duopolies.

In the case of monopolies, we did not consider hyperbolic price function, since in this case the monopoly profit would be

$$\frac{q}{q} - cq = 1 - cq$$

which is strictly decreasing in q, so (since at $q = 0$, this is not defined) the firm would select as small positive q values as possible. So this case is not interesting at all.

In the cases of mixed duopolies, one firm selects output level, while the other selects price as decision variable. Assume firm 1 is a Cournot firm and firm 2 is a Bertrand firm. We will discuss this case first in the linear model. The decision variable pair is (q_1, p_2) in this case. Solving (3.7) for q_2 and p_1 is easy. From the second equation

$$q_2 = \frac{1}{\beta_2}(\alpha_2 - \gamma_2 q_1 - p_2) \tag{3.18}$$

and then from the first equation

$$\begin{aligned} p_1 &= \alpha_1 - \beta_1 q_1 - \frac{\gamma_1}{\beta_2}(\alpha_2 - \gamma_2 q_1 - p_2) \\ &= \frac{1}{\beta_2}(\alpha_1\beta_2 - \alpha_2\gamma_1 - (\beta_1\beta_2 - \gamma_1\gamma_2)\, q_1 + \gamma_1 p_2)\,. \end{aligned} \tag{3.19}$$

Therefore, the profit functions of the two firms become

$$\begin{aligned} \pi_1'(q_1, p_2) &= q_1\left[\frac{1}{\beta_2}(\alpha_1\beta_2 - \alpha_2\gamma_1 - (\beta_1\beta_2 - \gamma_1\gamma_2)\, q_1 + \gamma_1 p_2) - c_1\right] \\ &= q_1\left(\frac{\alpha_1\beta_2 - \alpha_2\gamma_1}{\beta_2} - \frac{\beta_1\beta_2 - \gamma_1\gamma_2}{\beta_2} q_1 + \frac{\gamma_1}{\beta_2} p_2 - c_1\right) \end{aligned} \tag{3.20}$$

and

$$\pi_2'(q_1, p_2) = \frac{1}{\beta_2}(\alpha_2 - \gamma_2 q_1 - p_2)(p_2 - c_2)\,. \tag{3.21}$$

Notice that these profit functions have the same structures as the profits in the Cournot markets.

Consider next the hyperbolic case. Assume again that the first firm is Cournot and the second firm is Bertrand. Then, assuming q_1 and p_2 given, we solve equations (3.14) and (3.15) for q_2 and p_1:

$$q_2 = \frac{1 - p_2 q_1 \theta_2}{p_2} = \frac{1}{p_2} - q_1\theta_2 \tag{3.22}$$

and

$$p_1 = \frac{p_2}{(1 - \theta_1\theta_2) q_1 p_2 + \theta_1} \tag{3.23}$$

implying that the profit functions of the two firms are as follows:

$$\bar{\pi}_1'(q_1, p_2) = q_1 \left[\frac{p_2}{(1-\theta_1\theta_2)q_1 p_2 + \theta_1} - c_1 \right] \tag{3.24}$$

and

$$\bar{\pi}_2'(q_1, p_2) = \left(\frac{1}{p_2} - q_1\theta_2 \right) (p_2 - c_2) . \tag{3.25}$$

From (3.20) and (3.21), it is easy to see that

$$\frac{\partial \pi_1'(q_1, p_2)}{\partial q_1} = \frac{\alpha_1\beta_2 - \alpha_2\gamma_1}{\beta_2} - 2q_1 \frac{\beta_1\beta_2 - \gamma_1\gamma_2}{\beta_2} + \frac{\gamma_1}{\beta_2} p_2 - c_1 \tag{3.26}$$

so

$$\frac{\partial^2 \pi_1'(q_1, p_2)}{\partial q_1^2} = -\frac{2\left(\beta_1\beta_2 - \gamma_1\gamma_2\right)}{\beta_2} < 0$$

as $\beta_1\beta_2 - \gamma_1\gamma_2 > 0$, and

$$\frac{\partial \pi_2'(q_1, p_2)}{\partial p_2} = \frac{1}{\beta_2} (-2p_2 + c_2 + \alpha_2 - \gamma_2 q_1) \tag{3.27}$$

and

$$\frac{\partial^2 \pi_2'(q_1, p_2)}{\partial p_2^2} = -\frac{2}{\beta_2} < 0$$

implying that in both cases, the profit functions are strictly concave in the individual decision variables of the firms.

In the hyperbolic case, we have from (3.24) and (3.25) that

$$\frac{\partial \bar{\pi}_1'(q_1, p_2)}{\partial q_1} = \frac{\theta_1 p_2}{[(1-\theta_1\theta_2)q_1 p_2 + \theta_1]^2} - c_1 \tag{3.28}$$

with

$$\frac{\partial^2 \bar{\pi}_1'(q_1, p_2)}{\partial q_1^2} = -\frac{2\theta_1 p_2^2 (1-\theta_1\theta_2)}{[(1-\theta_1\theta_2)q_1 p_2 + \theta_1]^3} < 0$$

and

$$\frac{\partial \bar{\pi}_2'(q_1, p_2)}{\partial p_2} = \frac{c_2}{p_2^2} - q_1\theta_2 \tag{3.29}$$

with

$$\frac{\partial^2 \bar{\pi}_2'(q_1, p_2)}{\partial p_2^2} = -\frac{2c_2}{p_2^3} < 0,$$

so the profits are concave again in the own decision variables of the firms.

3.1 Discrete Time Models

Consider first the linear duopoly (3.5) and assume that the firms adjust their outputs proportionally to their marginal profits. If it is positive, then the interest of the firm is to increase production level. If it is negative, then the firm wants to decrease it and if marginal profit is zero, then the firm believes that its production level is on its maximal level, so does not want to change it. This idea can be formulated by the discrete system:

$$\begin{aligned} q_1(t+1) &= q_1(t) + K_1\left(\alpha_1 - 2\beta_1 q_1(t) - \gamma_1 q_2(t) - c_1\right) \\ q_2(t+1) &= q_2(t) + K_2\left(\alpha_2 - 2\beta_2 q_2(t) - \gamma_2 q_1(t) - c_2\right) \end{aligned} \tag{3.30}$$

where it is usually assumed that $0 < K_1, K_2 \leq 1$.

The Jacobian of this system has the form

$$\boldsymbol{J} = \begin{pmatrix} 1 - 2K_1\beta_1 & -K_1\gamma_1 \\ -K_2\gamma_2 & 1 - 2K_2\beta_2 \end{pmatrix}.$$

The characteristic polynomial can be written as

$$\varphi(\lambda) = \lambda^2 + a_1\lambda + a_2, \tag{3.31}$$

where the coefficients are

$$a_1 = 2\left(-1 + K_1\beta_1 + K_2\beta_2\right)$$

and

$$a_2 = 1 - 2K_1\beta_1 - 2K_2\beta_2 + K_1K_2\left(4\beta_1\beta_2 - \gamma_1\gamma_2\right).$$

Since system (3.30) is linear, local asymptotical stability implies global stability which occurs if and only if all roots of $\varphi(\lambda)$ are inside the unit circle. From Appendix C, we know that this is the case when

$$a_2 < 1,$$

$$a_1 + a_2 + 1 > 0,$$

$$-a_1 + a_2 + 1 > 0$$

each of which can be simplified as follows,

$$2K_1\beta_1 + 2K_2\beta_2 - K_1K_2\left(4\beta_1\beta_2 - \gamma_1\gamma_2\right) > 0,$$

$$K_1K_2\left(4\beta_1\beta_2 - \gamma_1\gamma_2\right) > 0,$$

$$4K_1\beta_1 + 4K_2\beta_2 - K_1K_2\left(4\beta_1\beta_2 - \gamma_1\gamma_2\right) < 4.$$

Under realistic assumption that $\beta_1\beta_2 - \gamma_1\gamma_2 > 0$, the second condition is satisfied. To illustrate the first and last conditions, we select $\beta_1 = \beta_2 = \gamma_1 = \gamma_2 = 1$, when we have

$$2K_1 + 2K_2 - 3K_1K_2 > 0,$$

$$4K_1 + 4K_2 - 3K_1K_2 < 4.$$

Both are hyperbolic inequalities:

$$\begin{aligned} K_2(2 - 3K_1) &> -2K_1, \\ K_2(4 - 3K_1) &< 4 - 4K_1. \end{aligned} \tag{3.32}$$

From the first inequality, we get

$$K_2 \begin{cases} > -\dfrac{2K_1}{2 - 3K_1} \text{ if } K_1 < \dfrac{2}{3}, \\ < -\dfrac{2K_1}{2 - 3K_1} \text{ if } K_1 > \dfrac{2}{3}, \end{cases} \tag{3.33}$$

where the first part is always satisfied since the right hand side is negative. The second condition of (3.32) holds if

$$K_2 < \frac{4 - 4K_1}{4 - 3K_1}. \tag{3.34}$$

It is easy to prove that as $K_1 > 2/3$,

$$\frac{4 - 4K_1}{4 - 3K_1} < \frac{2K_1}{3K_1 - 2},$$

so (3.34) is a stronger condition than the second part of (3.33).

Proposition 3.1 *The shaded area is the stability region in the (K_1, K_2) plane with $\beta_1 = \beta_2 = \gamma_1 = \gamma_2 = 1$ as shown in Fig. 3.1 .*

Assume next that there is a unit delay in the information of firm 1 about the output of the other firm. Then, system (3.30) is modified as follows:

$$\begin{aligned} q_1(t+1) &= q_1(t) + K_1\left(\alpha_1 - 2\beta_1 q_1(t) - \gamma_1 q_2(t-1) - c_1\right), \\ q_2(t+1) &= q_2(t) + K_2\left(\alpha_2 - 2\beta_2 q_2(t) - \gamma_2 q_1(t) - c_2\right). \end{aligned} \tag{3.35}$$

Fig. 3.1 Stability region in the (K_1, K_2) plane

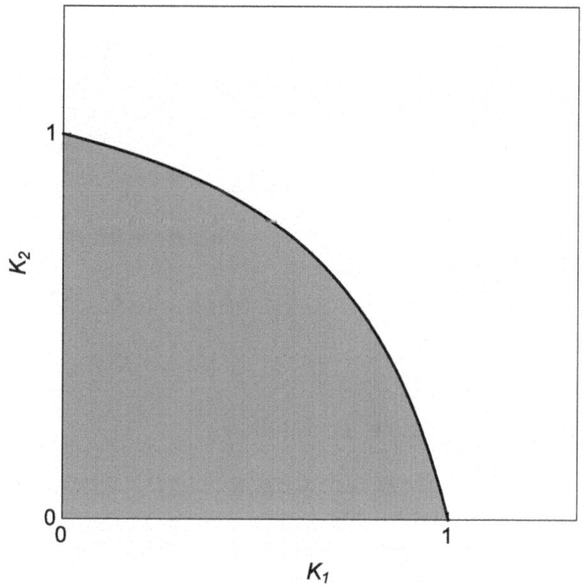

This delayed system can be rewritten as a three-dimensional no-delay system by introducing the new variable $z(t) = q_2(t-1)$:

$$\begin{aligned} q_1(t+1) &= q_1(t) + K_1(\alpha_1 - 2\beta_1 q_1(t) - \gamma_1 z(t) - c_1), \\ q_2(t+1) &= q_2(t) + K_2(\alpha_2 - 2\beta_2 q_2(t) - \gamma_2 q_1(t) - c_2), \\ z(t+1) &= q_2(t), \end{aligned} \tag{3.36}$$

with Jacobian matrix

$$\boldsymbol{J} = \begin{pmatrix} 1 - 2K_1\beta_1 & 0 & -K_1\gamma_1 \\ -K_2\gamma_2 & 1 - 2K_2\beta_2 & 0 \\ 0 & 1 & 0 \end{pmatrix}.$$

The characteristic polynomial is therefore

$$\begin{aligned} \varphi(\lambda) &= \det(\boldsymbol{J} - \lambda \boldsymbol{I}) \\ &= -\left(\lambda^3 + a_1\lambda^2 + a_2\lambda + a_3\right), \end{aligned} \tag{3.37}$$

where

$$a_1 = -2 + 2K_1\beta_1 + 2K_2\beta_2,$$

$$a_2 = 1 - 2K_1\beta_1 - 2K_2\beta_2 + 4K_1K_2\beta_1\beta_2,$$

and

$$a_3 = -K_1 K_2 \gamma_1 \gamma_2.$$

Appendix C gives sufficient and necessary conditions for all roots of $\varphi(\lambda)$ being inside the unit circle. These conditions are very complicated in general, so we assume again that $\beta_1 = \beta_2 = \gamma_1 = \gamma_2 = 1$, when

$$a_1 = 2K_1 + 2K_2 - 2,$$

$$a_2 = 4K_1 K_2 - 2K_1 - 2K_2 + 1$$

$$a_3 = -K_1 K_2.$$

Then, the conditions are as follows

$$1 + 2K_1 + 2K_2 - 2 + 4K_1 K_2 - 2K_1 - 2K_2 + 1 - K_1 K_2 > 0,$$

$$1 - 2K_1 - 2K_2 + 2 + 4K_1 K_2 - 2K_1 - 2K_2 + 1 + K_1 K_2 > 0,$$

$$1 - 4K_1 K_2 + 2K_1 + 2K_2 - 1 - (2K_1 + 2K_2 - 2)\, K_1 K_2 - (K_1 K_2)^2 > 0,$$

$$4K_1 K_2 - 2K_1 - 2K_2 + 1 < 3.$$

The first condition is reduced to $3K_1 K_2 > 0$, the second condition can be rewritten as $4(1 - K_1)(1 - K_2) + K_1 K_2 > 0$, and the last condition has the form $2\left(K_1 + K_2\left(1 - K_1\right) + \left(1 - K_1 K_2\right)\right) > 0$—they are all satisfied. The third condition is a quadratic inequality in K_2 with fixed values of K_1,

$$-K_2^2\left(2K_1 + K_1^2\right) + 2K_2\left(1 - K_1 - K_1^2\right) + 2K_1 > 0.$$

The discriminant D of the left hand side is positive, it has a negative root, and the positive root is

$$K_2^*(K_1) = \frac{2\left(1 - K_1 - K_1^2\right) + \sqrt{D}}{2\left(2K_1 + K_1^2\right)}.$$

This curve divides the (K_1, K_2) space into two parts. The stability region is under the curve, while the instability region is above the curve. To further simplify the conditions, we assume that $K_1 = K_2 = K$; then, we have the following reduced conditions,

$$3K^2 > 0,$$

$$4 - 8K + 5K^2 > 0,$$

$$4 - 2K - 4K^2 - K^3 > 0,$$

$$4K^2 - 4K < 2.$$

Since $0 < K \leq 1$, the first and last conditions hold. The discriminant in the second condition is $D = 64 - 80 < 0$, so it also holds for all K, so the stability condition is now reduced to

$$g(K) = K^3 + 4K^2 + 2K - 4 < 0. \tag{3.38}$$

Notice that $g(0) = -4,\ g(1) = 3 > 0$, and $g(K)$ is strictly increasing; therefore, there is a unique positive root $K^* = \sqrt{3} - 1 \simeq 0.732$ in $(0, 1)$, and (3.38) holds if $K < K^*$.

Proposition 3.2 *If* $\beta_1 = \beta_2 = \gamma_1 = \gamma_2 = 1$, *then system* (3.35) *is stable if* $K_2 < K_2^*(K_1)$. *If* $K_1 = K_2 = K$, *the stability condition is* $K < \sqrt{3} - 1 \simeq 0.732$.

Consider next the Bertrand duopoly, when model (3.30) is modified as

$$p_k(t+1) = p_k(t) + K_k\left[\frac{1}{\beta_1\beta_2 - \gamma_1\gamma_2}\left(\alpha_k\beta_\ell - \alpha_\ell\gamma_k - 2\beta_\ell p_k(t) + \gamma_k p_\ell(t) + c_k\beta_\ell\right)\right] \tag{3.39}$$

for $k = 1, 2$ with $\ell \neq k$. This is also a linear system similar to model (3.30), so its stability analysis is basically the same with different parameters. The case of mixed duopolies is also similar. These cases therefore are not discussed here in detail, they are left as simple exercises to the interested readers.

We will next investigate the hyperbolic case. Since the marginal profits are nonlinear functions of the decision variables of the firms, local linearization will be used around the equilibrium. Therefore, we need to find the equilibrium before linearization. The Cournot competition will be discussed first in detail.

The best response of firm k can be obtained from the first-order condition which makes the marginal profit zero. That is, from (3.13),

$$\frac{\theta_k q_\ell}{(q_k + \theta_k q_\ell)^2} - c_k = 0$$

implying that

$$\theta_k q_\ell = c_k\left(q_k + \theta_k q_\ell\right)^2\ (k = 1, 2)\,.$$

Dividing this equation with $k = 1$ by that with $k = 2$, we have

$$\frac{\theta_1 q_2}{\theta_2 q_1} = \frac{c_1}{c_2}\left(\frac{q_1 + \theta_1 q_2}{q_2 + \theta_2 q_1}\right)^2. \tag{3.40}$$

Introduce the variables,

$$c = \frac{c_2}{c_1} \text{ and } z = \frac{q_2}{q_1},$$

and then (3.40) implies that

$$c\frac{\theta_1}{\theta_2}z = \left(\frac{1+\theta_1 z}{z+\theta_2}\right)^2 \tag{3.41}$$

which is a cubic equation for z. Its solution is the intersection of the curves of

$$f(z) = c\frac{\theta_1}{\theta_2}z \text{ and } g(z) = \left(\frac{1+\theta_1 z}{z+\theta_2}\right)^2.$$

Clearly, $f(z)$ is a straight line with positive slope passing through the origin. In addition, notice that

$$g(0) = \left(\frac{1}{\theta_2}\right)^2, \; g'(z) < 0, \; g''(z) > 0 \text{ and } \lim_{z\to\infty} g(z) = \theta_1^2 > 0$$

implying that there is unique positive intersection z^* of the two curves which is clearly strictly decreasing in c. Since at the equilibrium $z^* = q_2^*/q_1^*$, from Eq. (3.40), we can derive the equilibrium output levels of the firms,

$$q_1^* = \frac{\theta_2}{c_2\left(\theta_2 + z^*\right)^2} \text{ and } q_2^* = \frac{\theta_2 z^*}{c_2\left(\theta_2 + z^*\right)^2}.$$

In our earlier models, the so-called *gradient adjustment* dynamics was assumed when the output adjustments were proportional to the marginal profits. In the literature of dynamic economic models, a different type of adjustment process is also examined, which is known as *best response* dynamics. In this case, the firms adjust their outputs toward their best responses. Equating the marginal profit (3.13) with zero, we have the best response functions of the firms as follows:

$$\begin{aligned} R_1(q_2) &= \sqrt{\frac{\theta_1 q_2}{c_1}} - \theta_1 q_2, \\ R_2(q_1) &= \sqrt{\frac{\theta_2 q_1}{c_2}} - \theta_2 q_1, \end{aligned} \tag{3.42}$$

and the best response dynamic equations take the forms:

$$
\begin{aligned}
q_1(t+1) &= q_1(t) + K_1\left(\sqrt{\frac{\theta_1 q_2(t)}{c_1}} - \theta_1 q_2(t) - q_1(t)\right) \\
q_2(t+1) &= q_2(t) + K_2\left(\sqrt{\frac{\theta_2 q_1(t)}{c_2}} - \theta_2 q_1(t) - q_2(t)\right)
\end{aligned}
\tag{3.43}
$$

where it is assumed again that $0 < K_1, K_2 \leq 1$. The Jacobian of this system is the following:

$$
J = \begin{pmatrix} 1-K_1 & K_1\left(\frac{1}{2}\sqrt{\frac{\theta_1}{c_1 q_2^*}} - \theta_1\right) \\ K_2\left(\frac{1}{2}\sqrt{\frac{\theta_2}{c_2 q_1^*}} - \theta_2\right) & 1-K_2 \end{pmatrix}.
$$

Let γ_1 and γ_2 denote the derivatives of the best response functions at the equilibrium, then the characteristic polynomial becomes

$$
\begin{aligned}
\varphi(\lambda) &= (1-K_1-\lambda)(1-K_2-\lambda) - K_1K_2\gamma_1\gamma_2 \\
&= \lambda^2 + (K_1+K_2-2)\,\lambda + (1-K_1)(1-K_2) - K_1K_2\gamma_1\gamma_2.
\end{aligned}
\tag{3.44}
$$

The system is locally asymptotically stable if all eigenvalues are inside the unit circle. Appendix C presents a sufficient and necessary condition:

$$
\begin{aligned}
K_1 + K_2 - 2 + (1-K_1)(1-K_2) - K_1K_2\gamma_1\gamma_2 + 1 &> 0, \\
-(K_1 + K_2 - 2) + (1-K_1)(1-K_2) - K_1K_2\gamma_1\gamma_2 + 1 &> 0, \\
(1-K_1)(1-K_2) - K_1K_2\gamma_1\gamma_2 &< 1,
\end{aligned}
$$

which can be rewritten as

$$
\begin{aligned}
K_1K_2\gamma_1\gamma_2 &< K_1K_2, \\
K_1K_2\gamma_1\gamma_2 &< 4 - 2K_1 - 2K_2 + K_1K_2, \\
K_1K_2\gamma_1\gamma_2 &> K_1K_2 - K_1 - K_2.
\end{aligned}
$$

Next, we prove that at the equilibrium

$$
\gamma_1\gamma_2 < \frac{1}{4}.
$$

Based on Eqs. (3.41) and (3.42), it is easy to see that at the equilibrium,

$$\gamma_1 = \frac{(z^*)^{-1} - \theta_1}{2} \text{ and } \gamma_2 = \frac{z^* - \theta_2}{2},$$

implying that

$$\gamma_1\gamma_2 = \frac{1}{4}\left[\theta_1\theta_2 + 1 - \left(z^*\theta_1 + \frac{\theta_2}{z^*}\right)\right].$$

Since $z^* + 1/z^* \geq 2$ for all positive values of z^*,

$$\gamma_1\gamma_2 \leq \frac{1}{4}\left[\theta_1\theta_2 + 1 - 2\min\{\theta_1, \theta_2\}\right] < \frac{1}{4}.$$

The last inequality follows from the simple observation that

$$\theta_1\theta_2 - 2\theta_1 = \theta_1(\theta_2 - 2) < 0$$

and

$$\theta_1\theta_2 - 2\theta_2 = \theta_2(\theta_1 - 2) < 0.$$

The first condition is obvious, and the second can be written as

$$K_1K_2(\gamma_1\gamma_2 - 1) < 4 - 2K_1 - 2K_2$$

where the left hand side is negative and the right hand side is nonnegative for $K_1, K_2 \leq 1$. The third condition has the simple form,

$$\gamma_1\gamma_2 > 1 - \frac{K_1 + K_2}{K_1K_2}. \tag{3.45}$$

This holds if $\gamma_1\gamma_2 \geq 0$, and if $\gamma_1\gamma_2 < 0$, when its absolute value has to be sufficiently small, below

$$\frac{1}{K_1} + \frac{1}{K_2} - 1.$$

Proposition 3.3 *The hyperbolic Cournot duopoly without delay is locally asymptotically stable if* (3.45) *holds.*

Assume next that firm 1 has a unit time delay in obtaining information about the output level of its competitor. Then, model (3.43) is modified as follows:

$$\begin{aligned}
q_1(t+1) &= q_1(t) + K_1\left(\sqrt{\frac{\theta_1 q_2(t-1)}{c_1}} - \theta_1 q_2(t-1) - q_1(t)\right), \\
q_2(t+1) &= q_2(t) + K_2\left(\sqrt{\frac{\theta_2 q_1(t)}{c_2}} - \theta_2 q_1(t) - q_2(t)\right).
\end{aligned} \tag{3.46}$$

By introducing the new variable $z(t) = q_2(t-1)$, a three-dimensional system is obtained:

$$q_1(t+1) = q_1(t) + K_1\left(\sqrt{\frac{\theta_1 z(t)}{c_1}} - \theta_1 z(t) - q_1(t)\right),$$

$$q_2(t+1) = q_2(t) + K_2\left(\sqrt{\frac{\theta_2 q_1(t)}{c_2}} - \theta_2 q_1(t) - q_2(t)\right),$$

$$z(t+1) = q_2(t)$$

with Jacobian matrix

$$\boldsymbol{J} = \begin{pmatrix} 1-K_1 & 0 & K_1\gamma_1 \\ K_2\gamma_2 & 1-K_2 & 0 \\ 0 & 1 & 0 \end{pmatrix}.$$

The eigenvalue equation can be written as

$$\varphi(\lambda) = \det\begin{pmatrix} 1-K_1-\lambda & 0 & K_1\gamma_1 \\ K_2\gamma_2 & 1-K_2-\lambda & 0 \\ 0 & 1 & -\lambda \end{pmatrix}$$

$$= -\lambda^3 + (2-K_1-K_2)\,\lambda^2 - (1-K_1)\,(1-K_2)\,\lambda + K_1K_2\gamma_1\gamma_2 = 0$$

or

$$\lambda^3 - (2-K_1-K_2)\,\lambda^2 + (1-K_1)\,(1-K_2)\,\lambda - K_1K_2\gamma_1\gamma_2 = 0.$$

The equilibrium is stable if all roots of this equation are inside the unit circle. Appendix C gives sufficient and necessary conditions which are, in this case,

$$1 + (-2+K_1+K_2) + (1-K_1)\,(1-K_2) - K_1K_2\gamma_1\gamma_2 > 0,$$

$$1 + (2-K_1-K_2) + (1-K_1)\,(1-K_2) + K_1K_2\gamma_1\gamma_2 > 0,$$

$$1 - (1-K_1)\,(1-K_2) + (2-K_1-K_2)\,K_1K_2\gamma_1\gamma_2 - (K_1K_2\gamma_1\gamma_2)^2 > 0,$$

$$(1-K_1)\,(1-K_2) < 3.$$

Since $\gamma_1\gamma_2 < 1/4$, the first condition is satisfied as well as the last inequality. The second condition gives

$$\gamma_1\gamma_2 > \frac{-4+2(K_1+K_2)-K_1K_2}{K_1K_2} = \gamma_{12}^* \tag{3.47}$$

where the right hand side is negative. The third condition is a quadratic inequality in $\gamma_1\gamma_2$,

$$(K_1K_2)^2(\gamma_1\gamma_2)^2 - K_1K_2(2-K_1-K_2)\gamma_1\gamma_2 - (K_1+K_2-K_1K_2) < 0. \tag{3.48}$$

Since the leading coefficient is positive and the constant term is negative, the left hand side has a negative root and a positive root,

$$\gamma_{12}^{\pm} = \frac{2-K_1-K_2 \pm \sqrt{4+(K_1-K_2)^2}}{2K_1K_2}.$$

So all conditions hold if

$$\max\{\gamma_{12}^{*}, \gamma_{12}^{-}\} < \gamma_1\gamma_2 < \gamma_{12}^{+}. \tag{3.49}$$

Proposition 3.4 *The equilibrium of the hyperbolic duopoly with unit delay model* (3.46) *is locally asymptotically stable if* (3.49) *holds.*

In the special case of $K_1 = K_2 = 1$ (which is called best response dynamics), $\gamma_{12}^{*} = \gamma_{12}^{-} = -1$ and $\gamma_{12}^{+} = 1$, so in this case the stability condition is

$$\gamma_1\gamma_2 \in \left(-1, \frac{1}{4}\right).$$

Consider next the hyperbolic Bertrand duopoly. Without delay, the dynamic equations (3.43) become

$$\begin{aligned} p_1(t+1) &= p_1(t) + K_1\left(\sqrt{\frac{c_1p_2(t)}{\theta_1}} - p_1(t)\right) \\ p_2(t+1) &= p_2(t) + K_2\left(\sqrt{\frac{c_2p_1(t)}{\theta_2}} - p_2(t)\right) \end{aligned} \tag{3.50}$$

since from the first-order conditions,

$$p_k = R_k(p_\ell) = \sqrt{\frac{c_kp_\ell}{\theta_k}}$$

based on the marginal profit (3.17). The equilibrium is the solution of equations

$$p_1 = \sqrt{\frac{c_1p_2}{\theta_1}} \text{ and } p_2 = \sqrt{\frac{c_2p_1}{\theta_2}},$$

which can be rewritten as

$$p_1^2\theta_1 = c_1 p_2 \text{ and } p_2^2\theta_2 = c_2 p_1.$$

From the second equation,

$$p_1 = \frac{p_2^2\theta_2}{c_2}$$

and by substituting it into the first equation, we have

$$\frac{p_2^4\theta_2^2\theta_1}{c_2^2} = c_1 p_2$$

so

$$p_2^* = \left(\frac{c_1 c_2^2}{\theta_2^2\theta_1}\right)^{\frac{1}{3}} \tag{3.51}$$

and similarly

$$p_1^* = \left(\frac{c_1^2 c_2}{\theta_1^2\theta_2}\right)^{\frac{1}{3}}. \tag{3.52}$$

The derivatives of the best response functions at the equilibrium are as follows:

$$\gamma_k = \frac{1}{2}\sqrt{\frac{c_k}{\theta_k p_\ell}} = \frac{1}{2}\left(\frac{c_k\theta_\ell}{c_\ell\theta_k}\right)^{\frac{1}{3}} \quad (k = 1, 2) \tag{3.53}$$

so

$$\gamma_1\gamma_2 = \frac{1}{2}\left(\frac{c_1\theta_2}{c_2\theta_1}\right)^{\frac{1}{3}} \frac{1}{2}\left(\frac{c_2\theta_1}{c_1\theta_2}\right)^{\frac{1}{3}} = \frac{1}{4}.$$

The Jacobian of system (3.50) has now the form

$$\boldsymbol{J} = \begin{pmatrix} 1 - K_1 & K_1\gamma_1 \\ K_2\gamma_2 & 1 - K_2 \end{pmatrix}$$

with characteristic equation

$$\varphi(\lambda) = \lambda^2 + (K_1 + K_2 - 2)\,\lambda + (1 - K_1)\,(1 - K_2) - K_1 K_2 \gamma_1\gamma_2 = 0$$

or

$$\varphi(\lambda) = \lambda^2 + (K_1 + K_2 - 2)\,\lambda + \left(1 - K_1 - K_2 + \frac{3}{4}K_1 K_2\right) = 0.$$

The roots are inside the unit circle if and only if

$$1 - K_1 - K_2 + \frac{3}{4}K_1 K_2 < 1,$$

$$(K_1 + K_2 - 2) + \left(1 - K_1 - K_2 + \frac{3}{4}K_1K_2\right) + 1 > 0,$$

$$-(K_1 + K_2 - 2) + \left(1 - K_1 - K_2 + \frac{3}{4}K_1K_2\right) + 1 > 0,$$

which can be rewritten as

$$\frac{3}{4}K_1K_2 < K_1 + K_2,$$

$$\frac{3}{4}K_1K_2 > 0,$$

$$\frac{3}{4}K_1K_2 > -4 + 2K_1 + 2K_2.$$

Since $0 < K_1, K_2 \leq 1$, all conditions are satisfied, so we have the following result.

Proposition 3.5 *The equilibrium of the hyperbolic Bertrand duopoly model* (3.50) *without delay is always locally asymptotically stable.*

Assume next that there is a unit delay for firm 1 in obtaining information about the output level of the competitor. Then, model (3.46) is modified accordingly with different best response functions. The Jacobian is the same as in the Cournot case with the only difference that γ_1 and γ_2 are now given by (3.53). The characteristic polynomial is also the same with $\gamma_1\gamma_2 = 1/4$. The stability condition (3.47) holds, since the right hand side is negative, and therefore condition (3.49) specializes as

$$\frac{1}{4} < \frac{2 - K_1 - K_2 + \sqrt{4 + (K_1 - K_2)^2}}{2K_1K_2}.$$

It is easy to see that the right hand side is always larger than $1/4$, so we have the following result.

Proposition 3.6 *The equilibrium of the hyperbolic Bertrand duopoly model with one delay is always locally asymptotically stable.*

The case of mixed hyperbolic duopolies is basically the same as the hyperbolic Cournot duopoly with different γ_1 and γ_2 values, where $\gamma_1\gamma_2 < 1/4$ as well, which will be proved later in the next section. Its continuous time scale variant also will be discussed there.

3.2 Continuous Time Models with Fixed Delays

For the sake of simplicity, we first discuss linear duopolies without product differentiation, when the common price function has the simple form,

$$p(q_1, q_2) = \alpha - \beta(q_1 + q_2) \tag{3.54}$$

and the profit of firm k is given as

$$\pi_k(q_1, q_2) = q_k\left[\alpha - \beta(q_k + q_\ell) - c_k\right] \tag{3.55}$$

with marginal profits

$$\frac{\partial \pi_k(q_1, q_2)}{\partial q_k} = \alpha - 2\beta q_k - \beta q_\ell - c_k, \ (k = 1, 2 \text{ and } \ell \neq k). \tag{3.56}$$

The best response of firm k can be obtained by equating the marginal profit to zero, when assuming interior optimum

$$q_k = \frac{\alpha - c_k}{2\beta} - \frac{q_\ell}{2}. \tag{3.57}$$

The equilibrium is obtained by solving equations

$$q_1 = \frac{\alpha - c_1}{2\beta} - \frac{q_2}{2} \text{ and } q_2 = \frac{\alpha - c_2}{2\beta} - \frac{q_1}{2}$$

to get

$$q_1^* = \frac{\alpha - 2c_1 + c_2}{3\beta} \text{ and } q_2^* = \frac{\alpha - 2c_2 + c_1}{3\beta}.$$

Similarly to the monopoly case, we assume that the rates of change in the outputs are proportional to the marginal profits of the firms leading to the following dynamic system,

$$\begin{aligned} \dot{q}_1(t) &= K_1 q_1(t)\left[\alpha - 2\beta q_1(t) - \beta q_2(t) - c_1\right], \\ \dot{q}_2(t) &= K_2 q_2(t)\left[\alpha - 2\beta q_2(t) - \beta q_1(t) - c_2\right]. \end{aligned} \tag{3.58}$$

The Jacobian of this system is obtained from the derivatives

$$\frac{\partial \dot{q}_1(t)}{\partial q_1(t)} = -2K_1\beta q_1^* \text{ and } \frac{\partial \dot{q}_2(t)}{\partial q_2(t)} = -2K_2\beta q_2^*$$

at the equilibrium, where we used the zero values of the marginal profits. Furthermore

$$\frac{\partial \dot{q}_1(t)}{\partial q_2(t)} = -K_1\beta q_1^* \text{ and } \frac{\partial \dot{q}_2(t)}{\partial q_1(t)} = -K_2\beta q_2^*.$$

So the Jacobian is the following:

$$J = \begin{pmatrix} -2K_1\beta q_1^* & -K_1\beta q_1^* \\ -K_2\beta q_2^* & -2K_2\beta q_2^* \end{pmatrix} = \begin{pmatrix} -2\beta\gamma_1 & -\beta\gamma_1 \\ -\beta\gamma_2 & -2\beta\gamma_2 \end{pmatrix}$$

where $\gamma_1 = K_1 q_1^*$ and $\gamma_2 = K_2 q_2^*$. In the case of interior equilibrium, both are positive. The characteristic polynomial can be written as

$$\varphi(\lambda) = \lambda^2 + (2\beta\gamma_1 + 2\beta\gamma_2)\lambda + 3\beta^2\gamma_1\gamma_2. \tag{3.59}$$

Since all coefficients are positive, the eigenvalues are negative or have negative real parts.

Proposition 3.7 *The equilibrium of the linear duopoly without product differentiation and without delay is locally asymptotically stable.*

Assume next that there is a delay in the information to firm 1 about the output of its competitor. Then, system (3.58) is modified as

$$\begin{aligned} \dot{q}_1(t) &= K_1 q_1(t)\left[\alpha - 2\beta q_1(t) - \beta q_2(t-\tau) - c_1\right], \\ \dot{q}_2(t) &= K_2 q_2(t)\left[\alpha - 2\beta q_2(t) - \beta q_1(t) - c_2\right], \end{aligned} \tag{3.60}$$

where $\tau > 0$ is the delay. Since at the equilibrium,

$$\frac{\partial \dot{q}_1(t)}{\partial q_1(t)} = -2\beta\gamma_1 \text{ and } \frac{\partial \dot{q}_2(t)}{\partial q_2(t)} = -2\beta\gamma_2$$

and

$$\frac{\partial \dot{q}_1(t)}{\partial q_2(t-\tau)} = -\beta\gamma_1 \text{ and } \frac{\partial \dot{q}_2(t)}{\partial q_1(t)} = -\beta\gamma_2,$$

the linearized system becomes

$$\begin{aligned} \dot{q}_{1\varepsilon}(t) &= -2\beta\gamma_1 q_{1\varepsilon}(t) - \beta\gamma_1 q_{2\varepsilon}(t-\tau), \\ \dot{q}_{2\varepsilon}(t) &= -\beta\gamma_2 q_{1\varepsilon}(t) - 2\beta\gamma_2 q_{2\varepsilon}(t). \end{aligned} \tag{3.61}$$

The characteristic polynomial can be obtained by substituting exponential solutions

$$q_{1\varepsilon}(t) = e^{\lambda t}u \text{ and } q_{2\varepsilon}(t) = e^{\lambda t}v$$

into system (3.61) to get

$$\begin{aligned} \lambda u &= -2\beta\gamma_1 u - \beta\gamma_1 e^{-\lambda\tau}v, \\ \lambda v &= -\beta\gamma_2 u - 2\beta\gamma_2 v \end{aligned}$$

and therefore

$$\varphi(\lambda) = \det \begin{pmatrix} -2\beta\gamma_1 - \lambda & -\beta\gamma_1 e^{-\lambda\tau} \\ -\beta\gamma_2 & -2\beta\gamma_2 - \lambda \end{pmatrix}$$
$$= \lambda^2 + 2\beta(\gamma_1 + \gamma_2)\lambda + \beta^2\gamma_1\gamma_2\left(4 - e^{-\lambda\tau}\right) = 0. \tag{3.62}$$

At $\tau = 0$, the equilibrium is stable, as we have already established it. As τ increases, stability might be lost, when $\lambda = iv$ $(v > 0)$. Substituting it into Eq. (3.62), we have

$$-v^2 + i2v\beta(\gamma_1 + \gamma_2) + \beta^2\gamma_1\gamma_2(4 - \cos v\tau + i\sin v\tau) = 0.$$

Separating the real and imaginary parts gives

$$-v^2 + 4\beta^2\gamma_1\gamma_2 = \beta^2\gamma_1\gamma_2\cos v\tau,$$

$$-2v\beta(\gamma_1 + \gamma_2) = \beta^2\gamma_1\gamma_2\sin v\tau.$$

Adding the squares of these equations, we see that

$$\left(-v^2 + 4\beta^2\gamma_1\gamma_2\right)^2 + (2v\beta(\gamma_1 + \gamma_2))^2 = \left(\beta^2\gamma_1\gamma_2\right)^2$$

or

$$v^4 + 4\beta^2\left(\gamma_1^2 + \gamma_2^2\right)v^2 + 15\beta^4\gamma_1^2\gamma_2^2 = 0. \tag{3.63}$$

Since this equation has no real roots, we have the following result.

Proposition 3.8 *The equilibrium in model* (3.60) *with one delay is locally asymptotically stable for all* $\tau > 0$.

Assume next that firm 1 has an average of two past output values of the competitor in forming its marginal profit. Then, equations (3.58) become

$$\dot{q}_1(t) = K_1q_1(t)\left[\alpha - 2\beta q_1(t) - \beta(\omega_1 q_2(t - \tau_1) + (1 - \omega_1)q_2(t - \tau_2)) - c_1\right],$$
$$\dot{q}_2(t) = K_2q_2(t)\left[\alpha - 2\beta q_2(t) - \beta q_1(t) - c_2\right], \tag{3.64}$$

where $\tau_1 > 0,\ \tau_2 > 0,\ \omega_1 \geq 0,\ \omega_2 \geq 0$ and $\omega_1 + \omega_2 = 1$. At the equilibrium,

$$\frac{\partial \dot{q}_1(t)}{\partial q_1(t)} = -2\beta\gamma_1,\ \frac{\partial \dot{q}_1(t)}{\partial q_2(t - \tau_1)} = -\beta\omega_1\gamma_1 \text{ and } \frac{\partial \dot{q}_1(t)}{\partial q_2(t - \tau_2)} = -\beta\omega_2\gamma_1$$

and the derivatives of $\dot{q}_2(t)$ are the same as before. So the linearized equations become

$$\dot{q}_{1\varepsilon}(t) = -2\beta\gamma_1 q_{1\varepsilon}(t) - \beta\omega_1\gamma_1 q_{2\varepsilon}(t - \tau_1) - \beta\omega_2\gamma_1 q_{2\varepsilon}(t - \tau_2),$$
$$\dot{q}_{2\varepsilon}(t) = -\beta\gamma_2 q_{1\varepsilon}(t) - 2\beta\gamma_2 q_{2\varepsilon}(t - \tau). \tag{3.65}$$

Substituting again the exponential forms

$$q_{1\varepsilon}(t) = e^{\lambda t}u \text{ and } q_{2\varepsilon}(t) = e^{\lambda t}v$$

into these equations, we can see that the characteristic equation is

$$\varphi(\lambda) = \det\begin{pmatrix} \lambda + 2\beta\gamma_1 & \beta\gamma_1\left(\omega_1 e^{-\lambda\tau_1} + \omega_2 e^{-\lambda\tau_2}\right) \\ \beta\gamma_2 & \lambda + 2\beta\gamma_2 \end{pmatrix}$$

$$= (\lambda + 2\beta\gamma_1)(\lambda + 2\beta\gamma_2) - \beta^2\gamma_1\gamma_2\left(\omega_1 e^{-\lambda\tau_1} + \omega_2 e^{-\lambda\tau_2}\right) = 0.$$

This equation is equivalent with the following:

$$1 + a_1(\lambda)e^{-\lambda\tau_1} + a_2(\lambda)e^{-\lambda\tau_2} = 0$$

with

$$a_k(\lambda) = \frac{-\beta^2\gamma_1\gamma_2}{(\lambda + 2\beta\gamma_1)(\lambda + 2\beta\gamma_2)}\omega_k \ (k = 1, 2). \tag{3.66}$$

We now follow the process outlined in Appendix B for two-delay equations. Notice first that

$$a_k(iv) = \frac{-\beta^2\gamma_1\gamma_2}{(iv + 2\beta\gamma_1)(iv + 2\beta\gamma_2)}\omega_k$$

and simple algebra shows that

$$|a_k(iv)| = \frac{\beta^2\gamma_1\gamma_2\omega_k}{\sqrt{\left(4\beta^2\gamma_1\gamma_2 - v^2\right)^2 + \left(2\beta(\gamma_1 + \gamma_2)v\right)^2}}. \tag{3.67}$$

Substituting this relation into (A11), its first condition for existence of solution requires that

$$v^4 + 4\beta^2\left(\gamma_1^2 + \gamma_2^2\right)v^2 + 15\left(\beta^2\gamma_1\gamma_2\right)^2 \le 0 \tag{3.68}$$

which cannot occur.

Proposition 3.9 *The equilibrium in model* (3.64) *with two delays is locally asymptotically stable with all positive values of* τ_1 *and* τ_2.

Assume next that firm 1 uses three past output values of the competitor in forming its marginal profit. Then, the first equation of (3.64) is modified as

$$\dot{q}_1(t) = K_1 q_1(t)\left[\alpha - 2\beta q_1(t) - \beta\left(\omega_1 q_2(t - \tau_1) + \omega_2 q_2(t - \tau_2) + \omega_3 q_2(t - \tau_3)\right) - c_1\right] \tag{3.69}$$

when the second equation has no change. Following the same procedure as it was shown in the case of two delays, we get the characteristic equation:

$$\varphi(\lambda) = (\lambda + 2\beta\gamma_1)(\lambda + 2\beta\gamma_2) - \beta^2\gamma_1\gamma_2\left(\omega_1 e^{-\lambda\tau_1} + \omega_2 e^{-\lambda\tau_2} + \omega_3 e^{-\lambda\tau_3}\right) = 0$$

which can be rewritten as

$$1 + a_1(\lambda)e^{-\lambda\tau_1} + a_2(\lambda)e^{-\lambda\tau_2} + a_3(\lambda)e^{-\lambda\tau_3} = 0$$

where $a_k(\lambda)$ is given by (3.66). Similarly to the two-delay case, triangular-like inequalities are the conditions for the existence of solution; one of them is

$$|a_1(iv)| + |a_2(iv)| + |a_3(iv)| \geq 1$$

which can be rewritten based on (3.67) as

$$\frac{\beta^2\gamma_1\gamma_2\sum_{k=1}^{3}\omega_k}{\sqrt{\left(4\beta^2\gamma_1\gamma_2 - v^2\right)^2 + \left(2\beta\left(\gamma_1+\gamma_2\right)v\right)^2}} \geq 1$$

leading to the same contradiction (3.68) as before.

Proposition 3.10 *The equilibrium in model* (3.69) *with three delays is locally asymptotically stable with all positive values of* τ_1, τ_2 *and* τ_3.

Up to this point, we considered a one-sided information delay, when firm 1 used one, two or even three past information on the output of its competitor to form its own marginal profit. Now, we assume that both firms have information delay, which can be modeled by the following delay equations:

$$\begin{aligned} \dot{q}_1(t) &= K_1 q_1(t)\left[\alpha - 2\beta q_1(t) - \beta q_2(t-\tau_1) - c_1\right] \\ \dot{q}_2(t) &= K_2 q_2(t)\left[\alpha - 2\beta q_2(t) - \beta q_1(t-\tau_2) - c_2\right] \end{aligned} \tag{3.70}$$

where $\tau_1 > 0$ and $\tau_2 > 0$ are the information delays. Notice that at the equilibrium,

$$\frac{\partial\dot{q}_1(t)}{\partial q_1(t)} = -2\beta K_1 q_1^* = -2\beta\gamma_1, \quad \frac{\partial\dot{q}_1(t)}{\partial q_2(t-\tau_1)} = -\beta K_1 q_1^* = -\beta\gamma_1,$$

$$\frac{\partial\dot{q}_2(t)}{\partial q_1(t-\tau_2)} = -\beta K_2 q_2^* = -\beta\gamma_2, \quad \frac{\partial\dot{q}_2(t)}{\partial q_2(t)} = -2\beta K_2 q_2^* = -2\beta\gamma_2,$$

so the linearized equations become

$$\dot{q}_{1\varepsilon}(t) = -2\beta\gamma_1 q_{1\varepsilon}(t) - \beta\gamma_1 q_{2\varepsilon}(t-\tau_1),$$

$$\dot{q}_{2\varepsilon}(t) = -\beta\gamma_2 q_{1\varepsilon}(t-\tau_2) - 2\beta\gamma_2 q_{2\varepsilon}(t).$$

Substituting the exponential solutions

$$q_{1\varepsilon}(t) = e^{\lambda t} u \text{ and } q_{2\varepsilon}(t) = e^{\lambda t} v$$

into these equations, we have

$$(\lambda + 2\beta\gamma_1)\, u + \beta\gamma_1 e^{-\lambda\tau_1} v = 0,$$

$$\beta\gamma_2 e^{-\lambda\tau_2} u + (\lambda + 2\beta\gamma_2)\, v = 0,$$

so the characteristic equation is the following:

$$\varphi(\lambda) = \det \begin{pmatrix} \lambda + 2\beta\gamma_1 & \beta\gamma_1 e^{-\lambda\tau_1} \\ \beta\gamma_2 e^{-\lambda\tau_2} & \lambda + 2\beta\gamma_2 \end{pmatrix}$$
$$= \lambda^2 + 2\beta\,(\gamma_1 + \gamma_2)\,\lambda + \beta^2\gamma_1\gamma_2 \left(4 - e^{-\lambda(\tau_1+\tau_2)}\right) = 0. \tag{3.71}$$

By introducing the new variable $\tau = \tau_1 + \tau_2$, this equation can be considered as the characteristic equation of a single-delay model. Without delays, $\tau = 0$ and the equilibrium is locally asymptotically stable as it has been shown earlier. As τ increases, stability might be lost, when $\lambda = iv$ $(v > 0)$. Substituting it into the characteristic equation, we have

$$-v^2 + 2\beta\,(\gamma_1 + \gamma_2)\, iv + 4\beta^2\gamma_1\gamma_2 - \beta^2\gamma_1\gamma_2\,(\cos v\tau - i \sin v\tau) = 0.$$

Separating the real and imaginary parts,

$$\beta^2\gamma_1\gamma_2 \cos v\tau = -v^2 + 4\beta^2\gamma_1\gamma_2,$$

$$\beta^2\gamma_1\gamma_2 \sin v\tau = -2\beta\,(\gamma_1 + \gamma_2)\, v.$$

By adding the squares of these equations, we have the following equation,

$$\beta^4\gamma_1^2\gamma_2^2 = \left(-v^2 + 4\beta^2\gamma_1\gamma_2\right)^2 + 4\beta^2\,(\gamma_1 + \gamma_2)^2\, v^2$$

or

$$v^4 + 4\beta^2\left(\gamma_1^2 + \gamma_2^2\right) v^2 + 15\beta^4\gamma_1^2\gamma_2^2 = 0.$$

This equation has no positive solution for v^2, implying that there is no stability switch.

Proposition 3.11 *The equilibrium of model* (3.70) *with two information delays is locally asymptotically stable with all positive values of* τ_1 *and* τ_2.

Assume next that both firms have implementation delays in their own output levels. The associated dynamic system is now the following:

$$\dot{q}_1(t) = K_1 q_1(t)\left[\alpha - 2\beta q_1(t-\tau_1) - \beta q_2(t) - c_1\right],$$
$$\dot{q}_2(t) = K_2 q_2(t)\left[\alpha - \beta q_1(t) - 2\beta q_2(t-\tau_2) - c_2\right]. \tag{3.72}$$

If only delayed values are known about the firms' own outputs, then their derivatives cannot be obtained. However, Eq. (3.72) can be solved numerically at $t = 0,\ t = h,\ t = 2h, \ldots$ with a small step size h, and at each step, only "older" output values are used. Similar to the previous case, it is easy to show that the characteristic polynomial is the following:

$$\varphi(\lambda) = \det\begin{pmatrix} -\lambda - 2\beta\gamma_1 e^{-\lambda\tau_1} & -\beta\gamma_1 \\ -\beta\gamma_2 & -\lambda - 2\beta\gamma_2 e^{-\lambda\tau_2} \end{pmatrix}$$
$$= \left(\lambda + 2\beta\gamma_1 e^{-\lambda\tau_1}\right)\left(\lambda + 2\beta\gamma_2 e^{-\lambda\tau_2}\right) - \beta^2\gamma_1\gamma_2 = 0$$

which can be rewritten as

$$P_0(\lambda) + P_1(\lambda)e^{-\lambda\tau_1} + P_2(\lambda)e^{-\lambda\tau_2} + P_3(\lambda)e^{-\lambda(\tau_1+\tau_2)} = 0$$

which is the same as (A1) in Appendix A where

$$P_0(\lambda) = \lambda^2 - \beta^2\gamma_1\gamma_2,$$
$$P_1(\lambda) = 2\beta\gamma_1\lambda,$$
$$P_2(\lambda) = 2\beta\gamma_2\lambda,$$
$$P_3(\lambda) = 4\beta^2\gamma_1\gamma_2.$$

We will now follow the method for three delays given in Appendix A.

Notice first that

$$P_0(iv) = -v^2 - \beta^2\gamma_1\gamma_2,$$
$$P_1(iv) = i2\beta\gamma_1 v,$$
$$P_2(iv) = i2\beta\gamma_2 v,$$
$$P_3(iv) = 4\beta^2\gamma_1\gamma_2.$$

so at $\lambda = iv$,

$$P_2\bar{P}_3 - P_0\bar{P}_1 = iv\left[-2v^2\beta\gamma_1 + \beta^3\gamma_1\gamma_2\left(8\gamma_2 - 2\gamma_1\right)\right].$$

Therefore

$$A_1(iv) = 0 \text{ and } B_1(iv) = v\left[-2v^2\beta\gamma_1 + \beta^3\gamma_1\gamma_2\left(8\gamma_2 - 2\gamma_1\right)\right].$$

Checking condition (A31), we can rewrite it as a polynomial inequality of degree 8:

$$P(v) = z^8 + 4a_3z^6 + 2a_2z^4 + 4a_1z^2 + a_0 \leq 0 \tag{3.73}$$

where $z = v/\beta$ and

$$a_3 = -2\gamma_1^2 + \gamma_1\gamma_2 - 2\gamma_2^2,$$

$$a_2 = 8\gamma_1^4 + 8\gamma_1^3\gamma_2 - 19\gamma_1^2\gamma_2^2 - 8\gamma_2^4 + 16\gamma_1^2\gamma_2\left(4\gamma_2 - \gamma_1\right),$$

$$a_1 = -30\gamma_1^4\gamma_2^2 - 15\gamma_1^3\gamma_2^3 + 30\gamma_1^2\gamma_2^4 - 4\gamma_1^2\gamma_2^2\left(4\gamma_2 - \gamma_1\right)^2,$$

$$a_0 = 225\gamma_1^4\gamma_2^4.$$

Introduce the new variable $Z = z^2$; then, we have relation

$$Z^4 + 4a_3Z^3 + 2a_2Z^2 + 4a_1Z + a_0 \leq 0.$$

The left hand side has four roots

$$Z_1 = 2\gamma_1^2 - \gamma_1\gamma_2 + 2\gamma_2^2 + 2\left|\gamma_1 - \gamma_2\right|\sqrt{\gamma_1^2 + \gamma_1\gamma_2 + \gamma_2^2},$$

$$Z_2 = 2\gamma_1^2 - \gamma_1\gamma_2 + 2\gamma_2^2 - 2\left|\gamma_1 - \gamma_2\right|\sqrt{\gamma_1^2 + \gamma_1\gamma_2 + \gamma_2^2},$$

$$Z_3 = 2\gamma_1^2 - \gamma_1\gamma_2 + 2\gamma_2^2 + 2\left|\gamma_1 - \gamma_2\right|\sqrt{\gamma_1^2 - 3\gamma_1\gamma_2 + \gamma_2^2},$$

$$Z_4 = 2\gamma_1^2 - \gamma_1\gamma_2 + 2\gamma_2^2 - 2\left|\gamma_1 - \gamma_2\right|\sqrt{\gamma_1^2 - 3\gamma_1\gamma_2 + \gamma_2^2}.$$

Since γ_1 and γ_2 are positive, the discriminant is positive in the first two cases. The discriminant of the last two cases is $D = \gamma_1^2 - 3\gamma_1\gamma_2 + \gamma_2^2$. If $D \geq 0$, then we have four real roots, and if $D < 0$, then two roots are real and two are complex. We have the following possibilities:

(a) If $\gamma_1 = \gamma_2$, then $0 < Z_1 = Z_2 = Z_3 = Z_4$;
(b) If $\gamma_1 \neq \gamma_2$ and $D \geq 0$, then $0 < Z_2 < Z_4 \leq Z_3 < Z_1$;
(c) If $\gamma_1 \neq \gamma_2$ and $D < 0$, then $0 < Z_2 < Z_1$.

We have now the following result.

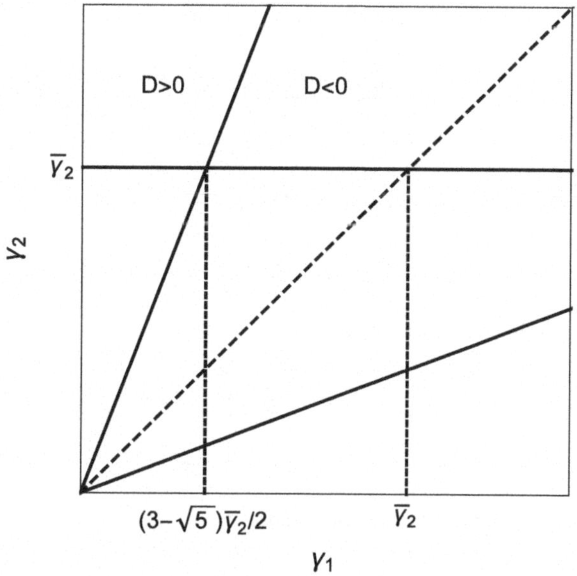

Fig. 3.2 Division of the (γ_1, γ_2) plane

Proposition 3.12 *Assume γ_1, γ_2 and β are positive.*

(*a*) *If $\gamma_1 = \gamma_2$, then $P(v) \geq 0$ for all $v > 0$,*

(*b*) *If $\gamma_1 \neq \gamma_2$ and $D \geq 0$, then $P(v) \leq 0$ for $v \in [v_2, v_4]$ and $v \in [v_3, v_1]$,*

(*c*) *If $\gamma_1 \neq \gamma_2$ and $D < 0$, then $P(v) \leq 0$ for $v \in [v_2, v_1]$,*

where $v_i = \beta\sqrt{Z_i}$ for $i = 1, 2, 3, 4$.

The value of the discriminant D becomes zero, when

$$\gamma_1 = \frac{3-\sqrt{5}}{2}\gamma_2 \text{ or } \gamma_1 = \frac{3+\sqrt{5}}{2}\gamma_2$$

which lines divide the (γ_1, γ_2) plane into three segments as it is shown in Fig. 3.2, where we assume that $\gamma_1 < \gamma_2$. If $\gamma_1 > \gamma_2$, then by interchanging the two firms, we get this condition, so it does not restrict generality.

The Bertrand and mixed linear cases can be examined in the same way, since the profit functions are also linear as in the above-discussed Cournot duopoly.

Next, we assume that both firms have information and implementation delays. In this case, model (3.70) is modified as follows:

$$\begin{aligned} \dot{q}_1(t) &= K_1 q_1(t)\left[\alpha - 2\beta q_1(t-\tau_1) - \beta q_2(t-\tau_1) - c_1\right] \\ \dot{q}_2(t) &= K_2 q_2(t)\left[\alpha - 2\beta q_2(t-\tau_2) - \beta q_1(t-\tau_2) - c_2\right]. \end{aligned} \tag{3.74}$$

The linearized equations have the forms

$$\dot{q}_{1\varepsilon}(t) = -2\beta\gamma_1 q_{1\varepsilon}(t-\tau_1) - \beta\gamma_1 q_{2\varepsilon}(t-\tau_1),$$

$$\dot{q}_{2\varepsilon}(t) = -\beta\gamma_2 q_{1\varepsilon}(t-\tau_2) - 2\beta\gamma_2 q_{2\varepsilon}(t-\tau_2).$$

Substituting the exponential forms

$$q_{1\varepsilon}(t) = e^{\lambda t}u \text{ and } q_{2\varepsilon}(t) = e^{\lambda t}v$$

into these equations, we have

$$\left(\lambda + 2\beta\gamma_1 e^{-\lambda\tau_1}\right)u + \beta\gamma_1 e^{-\lambda\tau_1}v = 0,$$

$$\beta\gamma_2 e^{-\lambda\tau_2}u + \left(\lambda + 2\beta\gamma_2 e^{-\lambda\tau_2}\right)v = 0,$$

giving the characteristic equation,

$$\varphi(\lambda) = \det\begin{pmatrix} \lambda + 2\beta\gamma_1 e^{-\lambda\tau_1} & \beta\gamma_1 e^{-\lambda\tau_1} \\ \beta\gamma_2 e^{-\lambda\tau_2} & \lambda + 2\beta\gamma_2 e^{-\lambda\tau_2} \end{pmatrix}$$
$$= P_0(\lambda) + P_1(\lambda)e^{-\lambda\tau_1} + P_2(\lambda)e^{-\lambda\tau_2} + P_3(\lambda)e^{-\lambda(\tau_1+\tau_2)} = 0 \tag{3.75}$$

where

$$P_0(\lambda) = \lambda^2,$$

$$P_1(\lambda) = 2\beta\gamma_1\lambda,$$

$$P_2(\lambda) = 2\beta\gamma_2\lambda,$$

$$P_3(\lambda) = 3\beta^2\gamma_1\gamma_2.$$

So this case is very similar to previous model and can be analyzed in an analogous manner.

Now, we turn our attention to the hyperbolic duopolies, when we will illustrate the dynamics with the mixed duopoly introduced in the introduction of this chapter. The first firm is quantity setting, and the second firm is price setting. The profit functions of the firms were given in Eqs. (3.24) and (3.25) with the corresponding marginal profits (3.28) and (3.29). In order to find the equilibrium, we first determine the best response functions of the firms. Equating (3.28) to zero and solving for q_1 gives

$$R_1(p_2) = \frac{1}{1-\theta_1\theta_2}\left(\sqrt{\frac{\theta_1}{c_1 p_2}} - \frac{\theta_1}{p_2}\right) \tag{3.76}$$

and from (3.29), we have

$$R_2(q_1) = \sqrt{\frac{c_2}{\theta_2 q_1}}. \tag{3.77}$$

The equilibrium is the solution of equations

$$q_1 = R_1(p_2) \text{ and } p_2 = R_2(q_1).$$

In order to solve these equations and use the trick again what we used for hyperbolic Cournot duopolies, we redefine the best responses in terms of quantities. From (3.14), (3.76) and (3.77), we have

$$\theta_1 (\theta_2 q_1 + q_2) = c_1 (q_1 + \theta_1 q_2)^2$$

and

$$\theta_2 q_1 = c_2 (\theta_2 q_1 + q_2)^2$$

implying that

$$c \frac{\theta_1}{\theta_2} (z + \theta_2) = \left(\frac{1 + \theta_1 z}{\theta_2 + z} \right)^2 \tag{3.78}$$

where $z = q_2/q_1$ and $c = c_2/c_1$ as before. If $g(z)$ denotes the right hand side, then

$$g(0) = \frac{1}{\theta_2^2} > 0, \; g'(0) < 0, \; g''(z) > 0 \text{ and } \lim_{z \to \infty} g(z) = \theta_1^2 > 0.$$

The left hand side is a linear function, strictly increasing with value of $c\theta_1 > 0$ at $z = 0$. A unique positive solution of equation (3.78) exists if and only if $c < 1/\theta_1 \theta_2^2$. Let the solution be denoted as z^*. Figure 3.3 shows the intersection of the left and right hand sides of equation (3.78), when $f(z)$ denotes the left hand side. The equilibrium of the mixed duopoly is given as

$$q_1^* = \frac{\theta_2}{c_2 (\theta_2 + z^*)^2} \text{ and } p_2^* = \frac{c_2 (\theta_2 + z^*)}{\theta_2}.$$

The best response dynamics of this mixed duopoly is derived by the modified equations of (3.43):

$$\begin{aligned} \dot{q}_1(t) &= K_1 \left[\frac{1}{1 - \theta_1 \theta_2} \left(\sqrt{\frac{\theta_1}{c_1 p_2(t)}} - \frac{\theta_1}{p_2(t)} \right) - q_1(t) \right], \\ \dot{p}_2(t) &= K_2 \left(\sqrt{\frac{c_2}{\theta_2 q_1(t)}} - p_2(t) \right), \end{aligned} \tag{3.79}$$

with the Jacobian at the equilibrium

Fig. 3.3 Determination of optimal output ratio

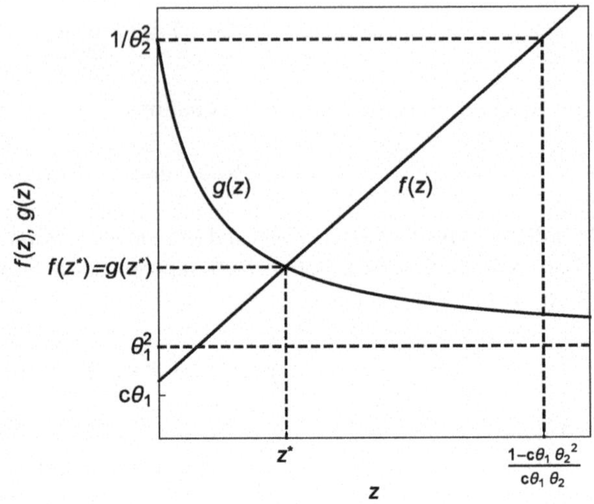

$$J = \begin{pmatrix} -K_1 & K_1\gamma_1 \\ K_2\gamma_2 & -K_2 \end{pmatrix}$$

where γ_k is the derivative of the best response function of firm k at the equilibrium,

$$\gamma_1 = \frac{\theta_1\theta_2^2}{(1-\theta_1\theta_2)\,c_2^2\,(\theta_2+z^*)^2}\left[1-\frac{1}{2}\sqrt{\frac{c\,(\theta_2+z^*)}{\theta_1\theta_2}}\right]$$

and

$$\gamma_2 = -\frac{c_2^2\,(\theta_2+z^*)^3}{2\theta_2^2}$$

with

$$\gamma_1\gamma_2 = \frac{1}{4} - \frac{\theta_1\,(z^*+\theta_2)}{4\,(1-\theta_1\theta_2)} < \frac{1}{4}$$

where Eq. (3.78) is used. The characteristic equation can be written as

$$\varphi(\lambda) = \lambda^2 + (K_1+K_2)\,\lambda + K_1K_2(1-\gamma_1\gamma_2) = 0. \tag{3.80}$$

Since all coefficients are positive, we have the following result.

Proposition 3.13 *The equilibrium of the mixed hyperbolic duopoly with model* (3.79) *is locally asymptotically stable.*

Assume next that both firms have delays in the competitor's variable. If τ_1 and τ_2 denote the delays, then the dynamic equations become

$$
\begin{aligned}
\dot{q}_1(t) &= K_1 \left[R_1 \left(p_2(t-\tau_1)\right) - q_1(t)\right], \\
\dot{p}_2(t) &= K_2 \left[R_2 \left(q_1(t-\tau_2)\right) - p_2(t)\right].
\end{aligned}
\tag{3.81}
$$

The Jacobian at the equilibrium is now given as

$$
\boldsymbol{J} = \begin{pmatrix} -K_1 - \lambda & K_1\gamma_1 e^{-\lambda\tau_1} \\ K_2\gamma_2 e^{\lambda\tau_2} & -K_2 - \lambda \end{pmatrix}
$$

with characteristic equation

$$
\varphi(\lambda) = \lambda^2 + (K_1 + K_2)\,\lambda + K_1 K_2 (1 - \gamma_1\gamma_2 e^{-\lambda\tau}) = 0, \tag{3.82}
$$

where we introduce the variable $\tau = \tau_1 + \tau_2$. So mathematically, we can deal with the problem as a single-delay case.

Notice first that $\lambda = 0$ cannot be an eigenvalue; otherwise, we would have $K_1 K_2 (1 - \gamma_1\gamma_2) = 0$ which is impossible, since $\gamma_1\gamma_2 < 1/4$. Without delay the equilibrium is locally asymptotically stable by Proposition 3.13. If τ becomes larger, then stability might be lost, when $\lambda = iv$ with some $v > 0$. Substituting $\lambda = iv$ into the characteristic equation,

$$
-v^2 + i\,(K_1 + K_2)\,v + K_1 K_2 \left[1 - \gamma_1\gamma_2 \left(\cos v\tau - i \sin v\tau\right)\right] = 0.
$$

Separating the real and imaginary parts, we have

$$
\begin{aligned}
-v^2 + K_1 K_2 \left(1 - \gamma_1\gamma_2 \cos v\tau\right) &= 0, \\
(K_1 + K_2)\,v + K_1 K_2 \gamma_1\gamma_2 \sin v\tau &= 0
\end{aligned}
$$

or

$$
\begin{aligned}
K_1 K_2 \gamma_1\gamma_2 \cos v\tau &= K_1 K_2 - v^2, \\
K_1 K_2 \gamma_1\gamma_2 \sin v\tau &= -\left(K_1 + K_2\right) v.
\end{aligned}
\tag{3.83}
$$

Adding the squares of these equations, a polynomial equation of degree 4 is obtained for v,

$$
v^4 + \left(K_1^2 + K_2^2\right) v^2 + (K_1 K_2)^2 \left[1 - (\gamma_1\gamma_2)^2\right] = 0.
$$

Let $z = v^2$, then this equation becomes quadratic:

$$
P(z) = z^2 + \left(K_1^2 + K_2^2\right) z + (K_1 K_2)^2 \left[1 - (\gamma_1\gamma_2)^2\right] = 0. \tag{3.84}
$$

If $\gamma_1\gamma_2 \geq -1$, then the left hand side is positive for all $z > 0$, implying the following result.

Proposition 3.14 *If* $\gamma_1\gamma_2 \geq -1$, *then the equilibrium of the mixed hyperbolic duopoly with the two-delay model* (3.81) *is locally asymptotically stable for all* $\tau_1,\ \tau_2 > 0$.

Consider now the case, when $\gamma_1\gamma_2 < -1$. The roots of equation (3.84) are

$$z_{1,2} = v_{1,2}^2 = \frac{-\left(K_1^2 + K_2^2\right) \pm \sqrt{\left(K_1^2 + K_2^2\right)^2 - 4\left(K_1 K_2\right)^2 \left[1 - \left(\gamma_1\gamma_2\right)^2\right]}}{2}$$

when one is positive and the other negative. Therefore, only one positive root exists for v, which is denoted by v^*. Then, the critical value of the delay is τ^* where

$$\cos v^*\tau^* = \frac{K_1 K_2 - v^{*2}}{K_1 K_2 \gamma_1\gamma_2} \text{ and } \sin v^*\tau^* = \frac{-\left(K_1 + K_2\right) v^*}{K_1 K_2 \gamma_1\gamma_2}$$

from (3.83). Since $\sin v^*\tau^* > 0$, clearly

$$\tau_n^* = \frac{1}{v^*}\left[\cos^{-1}\left(\frac{K_1 K_2 - v^{*2}}{K_1 K_2 \gamma_1\gamma_2}\right) + 2n\pi\right] \quad n = 0, 1, 2, \ldots \tag{3.85}$$

The direction of the stability switches at the critical values of τ_n^* is determined by using Hopf bifurcation as before. Assume that τ is the bifurcation parameter, and consider the eigenvalue as function of τ, $\lambda = \lambda(\tau)$. Implicitly differentiating the characteristic equation (3.82) with respect to τ, we have

$$2\lambda\lambda' + \left(K_1 + K_2\right)\lambda' + K_1 K_2 \left(-\gamma_1\gamma_2\right) e^{-\lambda\tau}\left(-\lambda'\tau - \lambda\right) = 0$$

implying that

$$\lambda' = -\frac{K_1 K_2 \gamma_1\gamma_2 \lambda e^{-\lambda\tau}}{2\lambda + \left(K_1 + K_2\right) + K_1 K_2 \gamma_1\gamma_2 \tau e^{-\lambda\tau}}.$$

From the characteristic equation, we know that

$$e^{-\lambda\tau} = \frac{\lambda^2 + \left(K_1 + K_2\right)\lambda + K_1 K_2}{K_1 K_2 \gamma_1\gamma_2}$$

and by substituting this expression into λ', after simple calculation, we can conclude that at $\lambda = iv^*$,

$$sign\left(\text{Re}\lambda'\right) = sign\left[\frac{\left(K_1 K_2 \gamma_1\gamma_2\right)^2}{K_1^2 + K_2^2 + 2v^{*2}}\right] > 0$$

showing that at each critical value τ_n^* at least one eigenvalue pair changes its real part from negative to positive.

Proposition 3.15 *If $\gamma_1\gamma_2 < -1$, then the equilibrium of the mixed hyperbolic duopoly with the two-delay model* (3.81) *is locally asymptotically stable with $\tau < \tau_0^*$ and unstable with $\tau > \tau_0^*$ and at $\tau = \tau_0^*$ Hopf bifurcation occurs giving the possibility of the birth of limit cycles.*

The condition $\gamma_1\gamma_2 < -1$ means that

$$\frac{(1 - \theta_1\theta_2) - \theta_1 z^* - \theta_1\theta_2}{4\,(1 - \theta_1\theta_2)} < -1$$

that is,

$$z^* > \frac{5 - 6\theta_1\theta_2}{\theta_1} = \bar{z}.$$

This is the case, when

$$f(\bar{z}) < g(\bar{z}).$$

The hyperbolic Cournot and Bertrand duopolies can be examined in the same way as shown before for mixed duopolies. Model (3.81) still applies when the appropriate best responses, $R_1(q_2)$, $R_2(q_1)$ or $R_1(p_2)$ and $R_2(p_1)$ are used, respectively. The characteristic polynomial (3.82) remains the same, where γ_1 and γ_2 are the derivatives of the corresponding best response functions at the equilibrium. In the Cournot case, $\gamma_1\gamma_2 < 1/4$, and in the Bertrand case, $\gamma_1\gamma_2 = 1/4$. Proposition 3.13 implies that the equilibrium of the hyperbolic Cournot and Bertrand duopolies without delays is locally asymptotically stable, and the same holds for the Bertrand model with two delays by Proposition 3.13. In case of the Cournot model, the equilibrium is locally asymptotically stable if $\gamma_1\gamma_2 \geq -1$; otherwise, Proposition 3.15 has to be applied.

3.3 Continuous Time Models with Distributed Delays

In this section, dynamic duopolies will be discussed, when the delays are continuously distributed. We start with a linear duopoly (3.54), but now we assume that the firms adjust their output proportionally with their marginal profits. Assume first that firm 1 has a delay in the output of its competitor. Then, the dynamic model has the following form

$$\begin{aligned} \dot{q}_1(t) &= K_1\left[\alpha - 2\beta q_1(t) - \beta q_2^e(t) - c_1\right], \\ \dot{q}_2(t) &= K_2\left[\alpha - 2\beta q_2(t) - \beta q_1(t) - c_2\right] \end{aligned} \tag{3.86}$$

where

$$q_2^e(t) = \int_0^t W(t-s, n, T) q_2(s) ds.$$

As it is introduced in Appendix B, the weighting function has the general form

$$W(t-s,n,T)=\begin{cases}\dfrac{1}{T}e^{-\frac{t-s}{T}} & \text{if } n=0,\\[2ex] \dfrac{1}{n!}\left(\dfrac{n}{T}\right)^{n+1}(t-s)^n e^{-\frac{n(t-s)}{T}} & \text{if } n\geq 1.\end{cases}$$

Since system (3.86) is linear, local asymptotic stability implies global asymptotic stability. The steady state of system (3.86) is the same as its steady state without delay. Let $q_{1\varepsilon}(t)$ and $q_{2\varepsilon}(t)$ denote the differences of $q_1(t)$ and $q_2(t)$ from their equilibrium levels. Then

$$\dot{q}_{1\varepsilon}(t)=K_1\left[-2\beta q_{1\varepsilon}(t)-\beta q^e_{2\varepsilon}(t)\right],$$

$$\dot{q}_{2\varepsilon}(t)=K_2\left[-2\beta q_{2\varepsilon}(t)-\beta q_{1\varepsilon}(t)\right].$$

The characteristic polynomial can be obtained by substituting the exponential forms

$$q_{1\varepsilon}(t)=e^{\lambda t}u \text{ and } q_{2\varepsilon}(t)=e^{\lambda t}v$$

into this homogeneous equation system. Then with

$$m=\begin{cases}1 \text{ if } n=0,\\ n \text{ if } n\geq 1,\end{cases}$$

as shown in Appendix B, we have

$$(\lambda+2K_1\beta)\,u+K_1\beta\left(1+\frac{\lambda T}{m}\right)^{-(n+1)}v=0,$$

$$K_2\beta u+(\lambda+2K_2\beta)\,v=0.$$

Nonzero solutions for u and v exist if the determinant of the system is zero, so the characteristic polynomial has the form

$$\varphi(\lambda)=\det\begin{pmatrix}\lambda+2K_1\beta & K_1\beta\left(1+\dfrac{\lambda T}{m}\right)^{-(n+1)}\\ K_2\beta & \lambda+2K_2\beta\end{pmatrix}$$

$$=(\lambda+2K_1\beta)(\lambda+2K_2\beta)-K_1K_2\beta^2\left(1+\frac{\lambda T}{m}\right)^{-(n+1)}=0.$$

We can rewrite this equation into a polynomial form,

$$(\lambda + 2K_1\beta)(\lambda + 2K_2\beta)\left(1 + \frac{\lambda T}{m}\right)^{n+1} - K_1K_2\beta^2 = 0. \tag{3.87}$$

Based on the special form of this equation, we can easily prove that all roots are either real negative numbers or their real parts are negative. Assume that for a solution $\text{Re}\lambda \geq 0$. Then

$$|\lambda + 2K_1\beta| \geq 2K_1\beta, \ |\lambda + 2K_2\beta| \geq 2K_2\beta \text{ and } \left|1 + \frac{\lambda T}{m}\right| \geq 1,$$

and therefore, the absolute value of the first term of (3.87) is at least $(2K_1\beta)(2K_2\beta) = 4K_1K_2\beta^2$, so the two terms cannot be equal.

Proposition 3.16 *The equilibrium with model* (3.86) *is asymptotically stable with all $n \geq 0$ and $T \geq 0$.*

The case of $T = 0$ corresponds to the no-delay system.

Assume next that firm 1 has delay in its own output. Then, system (3.86) is modified as

$$\begin{aligned} \dot{q}_1(t) &= K_1\left[\alpha - 2\beta q_1^e(t) - \beta q_2(t) - c_1\right], \\ \dot{q}_2(t) &= K_2\left[\alpha - 2\beta q_2(t) - \beta q_1(t) - c_2\right] \end{aligned} \tag{3.88}$$

where

$$q_1^e(t) = \int_0^t W(t - s, n, T) q_1(s) ds.$$

The characteristic polynomial is obtained from the following linear equations, which are derived by using exponential forms for the solutions

$$q_{1\varepsilon}(t) = e^{\lambda t} u \text{ and } q_{2\varepsilon}(t) = e^{\lambda t} v$$

as before,

$$\left(\lambda + 2K_1\beta\left(1 + \frac{\lambda T}{m}\right)^{-(n+1)}\right) u + K_1\beta v = 0,$$

$$K_2\beta u + (\lambda + 2K_2\beta) v = 0,$$

so the characteristic equation is as follows:

$$\varphi(\lambda) = \det\begin{pmatrix} \lambda + 2K_1\beta\left(1 + \frac{\lambda T}{m}\right)^{-(n+1)} & K_1\beta \\ K_2\beta & \lambda + 2K_2\beta \end{pmatrix}$$

$$= \left(\lambda + 2K_1\beta\left(1 + \frac{\lambda T}{m}\right)^{-(n+1)}\right)(\lambda + 2K_2\beta) - K_1K_2\beta^2 = 0,$$

or in polynomial form

$$\left(\lambda\left(1+\frac{\lambda T}{m}\right)^{n+1}+2K_1\beta\right)(\lambda+2K_2\beta)-K_1K_2\beta^2\left(1+\frac{\lambda T}{m}\right)^{n+1}$$
$$=\left(\lambda^2+2K_2\beta\lambda-K_1K_2\beta^2\right)\left(1+\frac{\lambda T}{m}\right)^{n+1}+2K_1\beta(\lambda+2K_2\beta)=0. \tag{3.89}$$

For large values of n, this is a large-degree polynomial. For the sake of simplicity, we consider the simple special case of $n=0$. Then, (3.89) is a cubic polynomial,

$$a_3\lambda^3+a_2\lambda^2+a_1\lambda+a_0=0$$

where

$$a_3=T,\ a_2=1+2K_2\beta T,\ a_1=2K_2\beta-K_1K_2\beta^2T+2K_1\beta \text{ and } a_0=3K_1K_2\beta^2.$$

The coefficient a_1 is positive if

$$T<\frac{2(K_1+K_2)}{K_1K_2\beta}=T^*. \tag{3.90}$$

From Theorem B.1, we know that the stability condition is $a_1a_2>a_0a_3$, which is now the following,

$$(1+2K_2\beta T)\left(2\beta(K_1+K_2)-K_1K_2\beta^2T\right)>3K_1K_2T\beta^2.$$

This is a quadratic inequality in T,

$$2K_1K_2^2\beta^2T^2-4K_2^2\beta T-2(K_1+K_2)<0. \tag{3.91}$$

The left hand side has a positive and negative real roots, the positive is

$$T^{**}=\frac{K_2+\sqrt{K_2^2+K_1(K_1+K_2)}}{K_1K_2\beta}$$

and (3.91) holds if $T<T^{**}$. However, it is easy to see that $T^{**}<T^*$, so we have the following result.

Proposition 3.17 *The equilibrium with model* (3.88) *is asymptotically stable if* $T<T^{**}$.

We will next introduce multiple delays into the dynamic models; however, instead of examining the effect of fixed delays on the stability of the equilibrium, continuously distributed delays will be considered. It has been proved in Proposition 3.7 that without delay the equilibrium is locally asymptotically stable. Interesting stability conditions were observed when the firms had implementation delays in their own output levels with model (3.72) when fixed delays were assumed. For the sake of simplicity, we assume again that the output level changes are proportional to the marginal profits, so we will examine system

$$\begin{aligned}\dot{q}_1(t) &= K_1\left[\alpha - c_1 - 2\beta q_1^e(t) - \beta q_2(t)\right],\\ \dot{q}_2(t) &= K_2\left[\alpha - c_2 - \beta q_1(t) - 2\beta q_2^e(t)\right],\end{aligned} \tag{3.92}$$

where

$$q_1^e(t) = \int_0^t W(t-s, n_1, T_1) q_1(s) ds$$

and

$$q_2^e(t) = \int_0^t W(t-s, n_2, T_2) q_2(s) ds.$$

Here, the weighting function is as before,

$$W(t-s, n, T) = \begin{cases} \dfrac{1}{T} e^{-\frac{t-s}{T}} & \text{if } n = 0,\\[2ex] \dfrac{1}{n!}\left(\dfrac{n}{T}\right)^{n+1} (t-s)^n e^{-\frac{n(t-s)}{T}} & \text{if } n \geq 1,\end{cases}$$

as it is given as Eq. (B1) in Appendix B.

Since system (3.92) is linear, local asymptotic stability implies global asymptotic stability. The steady state of this system is the same as that of the system without delays. Let $q_{1\varepsilon}$ and $q_{2\varepsilon}$ be the differences of $q_1(t)$ and $q_2(t)$ from their equilibrium levels, then the corresponding homogeneous system becomes

$$\begin{aligned}\dot{q}_{1\varepsilon}(t) &= K_1\left[-2\beta \textstyle\int_0^t W(t-s, n_1, T_1) q_{1\varepsilon}(s) ds - \beta q_{2\varepsilon}(t)\right],\\ \dot{q}_{2\varepsilon}(t) &= K_2\left[-\beta q_{1\varepsilon}(t) - 2\beta \textstyle\int_0^t W(t-s, n_2, T_2) q_{2\varepsilon}(s) ds\right].\end{aligned} \tag{3.93}$$

The characteristic polynomial can be obtained by substituting the exponential solutions,

$$q_{1\varepsilon}(t) = e^{\lambda t} u \text{ and } q_{2\varepsilon}(t) = e^{\lambda t} v,$$

into this system and using relations (B4) or (B5) from Appendix B as $t \to \infty$,

$$\left(\lambda + 2K_1\beta\left(1+\frac{\lambda T_1}{m_1}\right)^{-(n_1+1)}\right)u + K_1\beta v = 0,$$

$$K_2\beta u + \left(\lambda + 2K_2\beta\left(1+\frac{\lambda T_2}{m_2}\right)^{-(n_2+1)}\right)v = 0,$$

where for $k = 1, 2$, as before

$$m_k = \begin{cases} 1 & \text{if } n_k = 0, \\ n_k & \text{if } n_k \geq 1. \end{cases}$$

So the characteristic equation is given as follows

$$\begin{aligned} \varphi(\lambda) &= \det\begin{pmatrix} \lambda + 2K_1\beta\left(1+\frac{\lambda T_1}{m_1}\right)^{-(n_1+1)} & K_1\beta \\ K_2\beta & \lambda + 2K_2\beta\left(1+\frac{\lambda T_2}{m_2}\right)^{-(n_2+1)} \end{pmatrix} \\ &= \left(\lambda + 2K_1\beta\left(1+\frac{\lambda T_1}{m_1}\right)^{-(n_1+1)}\right)\left(\lambda + 2K_2\beta\left(1+\frac{\lambda T_2}{m_2}\right)^{-(n_2+1)}\right) - K_1K_2\beta^2 \\ &= 0. \end{aligned} \tag{3.94}$$

In order to have a polynomial equation, we multiply both sides by

$$\left(1+\frac{\lambda T_1}{m_1}\right)^{n_1+1} \quad \text{and} \quad \left(1+\frac{\lambda T_2}{m_2}\right)^{n_2+1}$$

to have

$$\left[\lambda\left(1+\frac{\lambda T_1}{m_1}\right)^{n_1+1} + 2K_1\beta\right]\left[\lambda\left(1+\frac{\lambda T_2}{m_2}\right)^{n_2+1} + 2K_2\beta\right]$$
$$- K_1K_2\beta^2\left(1+\frac{\lambda T_1}{m_1}\right)^{n_1+1}\left(1+\frac{\lambda T_2}{m_2}\right)^{n_2+1} = 0.$$

This is a very complicated high-degree polynomial equation, so we will examine only the most simple case of $n_1 = n_2 = 0$. Then, this is a fourth degree polynomial equation

$$a_0\lambda^4 + a_1\lambda^3 + a_2\lambda^2 + a_3\lambda + a_4 = 0$$

where

$$a_0 = T_1 T_2 > 0,$$

$$a_1 = T_1 + T_2 > 0,$$

$$a_2 = 1 + 2\beta\,(K_1 T_2 + K_2 T_1) - K_1 K_2 T_1 T_2 \beta^2,$$

$$a_3 = 2\,(K_1 + K_2)\,\beta - K_1 K_2\,(T_1 + T_2)\,\beta^2,$$

$$a_4 = 3K_1 K_2 \beta^2 > 0.$$

Notice that the signs of a_2 and a_3 are indeterminate. Appendix B presents sufficient and necessary condition to guarantee that all roots of this equation are negative or have negative real parts. The Hurwitz matrix is now

$$\boldsymbol{H}_4 = \begin{pmatrix} a_1 & a_0 & 0 & 0 \\ a_3 & a_2 & a_1 & a_0 \\ 0 & a_4 & a_3 & a_2 \\ 0 & 0 & 0 & a_4 \end{pmatrix}.$$

Since $a_0 > 0$ with principal minors

$$\boldsymbol{H}_1 = a_1 > 0,$$

$$\det(\boldsymbol{H}_2) = a_1 a_2 - a_0 a_3 = T_1 + T_2 + 2\beta\left(K_1 T_2^2 + K_2 T_1^2\right) > 0,$$

$$\det(\boldsymbol{H}_3) = a_3 \det(\boldsymbol{H}_2) - a_1^2 a_4,$$

$$\det(\boldsymbol{H}_4) = a_4 \det(\boldsymbol{H}_3).$$

Notice that if $\det(\boldsymbol{H}_3) > 0$, then $a_3 > 0$, and since $\det(\boldsymbol{H}_2) > 0$, a_2 has to be positive. Therefore, the only stability condition we have to check is the positivity of $\det(\boldsymbol{H}_3)$.

Proposition 3.18 *If $T_1, T_2 > 0$ and $n_1 = n_2 = 0$, then the continuously distributed delay system* (3.93) *is globally asymptotically stable for* (T_1, T_2) *under the partition curve* $\det(\boldsymbol{H}_3) = 0$ *and unstable above it, where*

$$\det(\boldsymbol{H}_3) = \beta\left\{\left[T_1 + T_2 + 2\beta\left(K_1 T_2^2 + K_2 T_1^2\right)\right]\left[2(K_1 + K_2) - K_1 K_2\,(T_1 + T_2)\,\beta\right] - 3\beta K_1 K_2\,(T_1 + T_2)^2\right\}.$$

The stability region is the shaded area in Fig. 3.4 with $\beta = K_1 = K_2 = 1$. Notice that at $T_2 = 0$,

$$\det(\boldsymbol{H}_3) = \beta\left\{\left[T_1 + 2\beta\left(K_2 T_1^2\right)\right]\left[2(K_1 + K_2) - K_1 K_2 T_1 \beta\right] - 3\beta K_1 K_2 T_1^2\right\}.$$

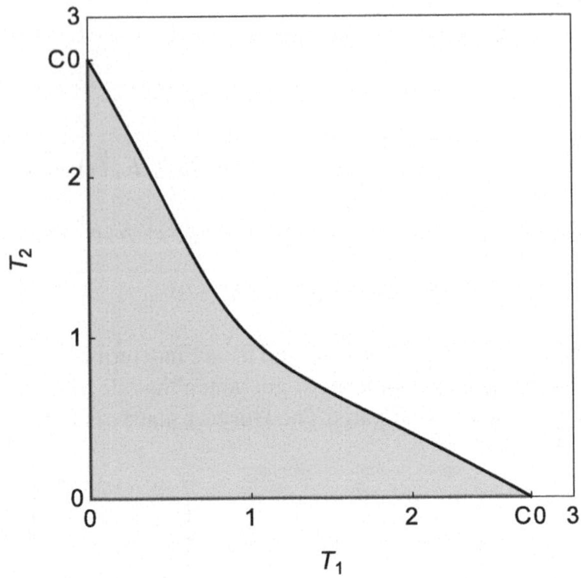

Fig. 3.4 Partition curve of continuous time model with distributed delay and $n_1 = n_2 = 0$

Clearly, $T_1 = 0$ is a root of $\det(\boldsymbol{H}_3) = 0$, and if $T_1 \neq 0$, then we get a quadratic equation

$$-\left(2K_1K_2^2\beta^2\right)T_1^2 + 4K_2^2\beta T_1 + 2(K_1 + K_2) = 0$$

where the quadratic coefficient is negative, the constant term is positive, and the linear coefficient is $4\beta K_2^2 > 0$. Therefore, one root is negative, the other is positive denoted by $C0$ in Fig. 3.4. If $T_1 = 0$, then similarly, we can see that $T_2 = 0$ is a root, and the other root is also $C0$ because of the symmetry between the firms.

The cases of Bertrand and mixed oligopolies are very similar, since in both cases the profit functions are linear. Therefore, their asymptotical analysis is the same as for the case shown before with slight modifications.

We turn our attention to the hyperbolic duopolies, and as our first case, we revisit model (3.86) with best response dynamics, which now has the form

$$\begin{aligned} \dot{q}_1(t) &= K_1\left[R_1\left(q_2^e(t)\right) - q_1(t)\right], \\ \dot{q}_2(t) &= K_2\left[R_2\left(q_1(t)\right) - q_2(t)\right], \end{aligned} \tag{3.95}$$

where

$$q_2^e(t) = \int_0^t W(t-s, n, T)q_2(s)ds.$$

The linearized model can be written as

$$\dot{q}_{1\varepsilon}(t) = K_1 \left[\gamma_1 \int_0^t W(t-s, n, T) q_{2\varepsilon}(s) ds - q_{1\varepsilon}(t) \right],$$

$$\dot{q}_{2\varepsilon}(t) = K_2 \left[\gamma_2 q_{1\varepsilon}(t) - q_{2\varepsilon}(t) \right],$$

where γ_1 and γ_2 are the derivatives of the best response functions at the equilibrium. By substituting the exponential solution forms

$$q_{1\varepsilon}(t) = e^{\lambda t} u \text{ and } q_{2\varepsilon}(t) = e^{\lambda t} v,$$

and letting $t \to \infty$, we have

$$(\lambda + K_1) u - K_1 \gamma_1 \left(1 + \frac{\lambda T}{m}\right)^{-(n+1)} v = 0,$$

$$-K_2 \gamma_2 u + (\lambda + K_2) v = 0,$$

implying that the characteristic equation becomes

$$\varphi(\lambda) = \det \begin{pmatrix} \lambda + K_1 & -K_1 \gamma_1 \left(1 + \frac{\lambda T}{m}\right)^{-(n+1)} \\ -K_2 \gamma_2 & \lambda + K_2 \end{pmatrix}$$

$$= (\lambda + K_1)(\lambda + K_2) - K_1 K_2 \gamma_1 \gamma_2 \left(1 + \frac{\lambda T}{m}\right)^{-(n+1)} = 0,$$

where as before

$$m = \begin{cases} 1 \text{ if } n = 0, \\ n \text{ if } n \geq 1. \end{cases}$$

The polynomial form of this equation is as follows:

$$(\lambda + K_1)(\lambda + K_2) \left(1 + \frac{\lambda T}{m}\right)^{n+1} - K_1 K_2 \gamma_1 \gamma_2 = 0. \tag{3.96}$$

Notice that if $\text{Re}\, \lambda \geq 0$, then

$$|\lambda + K_1| \geq K_1, \ |\lambda + K_2| \geq K_2 \text{ and } \left|1 + \frac{\lambda T}{m}\right| \geq 1,$$

so the absolute value of the first term is at least $K_1 K_2$. We have now the following result.

Proposition 3.19 *If* $|\gamma_1 \gamma_2| \leq 1$, *then the equilibrium with model* (3.95) *is asymptotically stable.*

We know that $\gamma_1\gamma_2 < 1/4$, so this result holds for $\gamma_1\gamma_2 \geq -1$. The case of $\gamma_1\gamma_2 < -1$ can be examined similarly to Proposition 3.15.

Consider next the special case of $n = 0$, when equation (3.96) is cubic:

$$\begin{aligned}&(\lambda + K_1)(\lambda + K_2)(1 + \lambda T) - K_1 K_2 \gamma_1 \gamma_2 \\ &= T\lambda^3 + (1 + T(K_1 + K_2))\lambda^2 + (K_1 + K_2 + TK_1K_2)\lambda + K_1K_2(1 - \gamma_1\gamma_2) = 0.\end{aligned}$$

We already established in Sect. 3.1 that $\gamma_1\gamma_2 < 1/4$, so all coefficients are positive. Due to Theorem B.1 in Appendix B, the eigenvalues have negative real parts if and only if

$$(1 + T(K_1 + K_2))(K_1 + K_2 + TK_1K_2) > K_1K_2(1 - \gamma_1\gamma_2)T$$

which can be rewritten as

$$\gamma_1\gamma_2 > -\frac{(K_1 + K_2)(1 + TK_1)(1 + TK_2)}{TK_1K_2} = \gamma_{12}^*.$$

We can also show that at the critical value of $\gamma_1\gamma_2 = \gamma_{12}^*$, Hopf bifurcation occurs with the birth of limit cycle around the equilibrium. Simple calculation shows that

$$z^* = 1 - \gamma_{12}^* = \frac{(1 + T(K_1 + K_2))(K_1 + K_2 + TK_1K_2)}{TK_1K_2} \tag{3.97}$$

and the cubic characteristic equation can be factored as

$$\left(\lambda + \frac{1 + T(K_1 + K_2)}{T}\right)\left(T\lambda^2 + (K_1 + K_2 + TK_1K_2)\right) = 0$$

showing that there is a negative real root,

$$\lambda_1 = -\frac{1 + T(K_1 + K_2)}{T}$$

and a pair of pure complex roots,

$$\lambda_{2,3} = \pm i\sqrt{\frac{K_1 + K_2 + TK_1K_2}{T}} = \pm iv.$$

We can now select $z = 1 - \gamma_1\gamma_2$ as bifurcation parameter and consider the eigenvalue as a function of z, $\lambda = \lambda(z)$. Implicit differentiation of the characteristic equation with respect to z shows that

$$\lambda' = -\frac{K_1K_2}{3T\lambda^2 + 2(1 + T(K_1 + K_2))\lambda + (K_1 + K_2 + TK_1K_2)}.$$

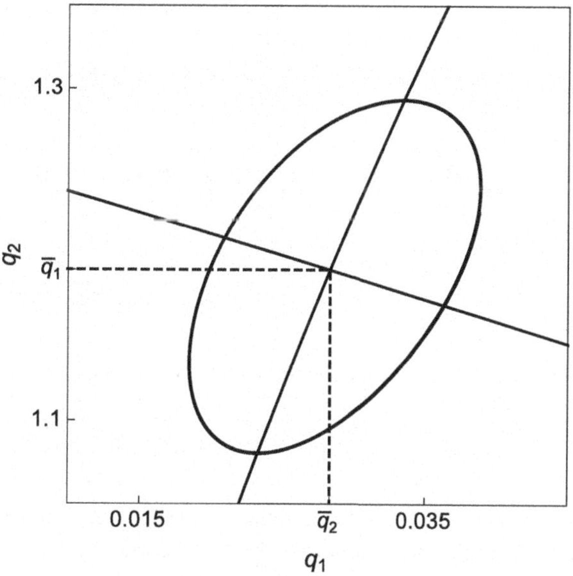

Fig. 3.5 Birth of limit cycle

At $\lambda = iv$,

$$\mathrm{Re}\lambda' = -\frac{K_1K_2\left(3T\lambda^2 + K_1 + K_2 + TK_1K_2\right)}{\left(3T\lambda^2 + K_1 + K_2 + TK_1K_2\right)^2 + (2v)^2\left(1 + T\left(K_1 + K_2\right)^2\right)}. \tag{3.98}$$

The denominator is positive, so the sign of $\mathrm{Re}\lambda'$ is the same as that of the numerator, which is, at the critical value,

$$2K_1K_2\left(K_1 + K_2 + TK_1K_2\right) > 0$$

showing that at least one pair of eigenvalues changes its real part from negative to positive, implying that at the critical value, stability is lost. A limit cycle is illustrated in Fig. 3.5 with the parameter selection,

$$K_1 = K_2 = 0.8,\ \theta_1 = 0.7915,\ \theta_2 = 0.5,\ c_1 = 1 \text{ and } c_2 = 0.00975.$$

As it is shown in Appendix B, system (3.95) with $n = 0$ can be rewritten as a three-dimensional ordinary differential equation, which is solved by standard numerical procedure.

In introducing multiple delays, assume now that each firm has delay in the information about the output of its competitor. If the delays are continuously distributed, then model (3.95) becomes

$$\begin{aligned}\dot{q}_1(t) &= K_1\left[R_1\left(q_2^e(t)\right) - q_1(t)\right], \\ \dot{q}_2(t) &= K_2\left[R_2\left(q_1^e(t)\right) - q_2(t)\right],\end{aligned} \tag{3.99}$$

with

$$q_2^e(t) = \int_0^t W(t-s, T_1)q_2(s)ds$$

and

$$q_1^e(t) = \int_0^t W(t-s, T_2)q_1(s)ds$$

where exponential weighting function (i.e., $n_1 = n_2 = 0$) is assumed for the sake of simplicity,

$$W(t-s, T) = \frac{1}{T}e^{-\frac{t-s}{T}}.$$

Let γ_1 and γ_2 denote the derivatives of the best response functions at the equilibrium, then the linearized equations have the form

$$\begin{aligned}\dot{q}_1(t) &= K_1\left[\gamma_1 \int_0^t W(t-s, T_1)q_2(s)ds - q_1(t)\right], \\ \dot{q}_2(t) &= K_2\left[\gamma_2 \int_0^t W(t-s, T_2)q_1(s)ds - q_2(t)\right].\end{aligned} \tag{3.100}$$

Substitute exponential forms

$$q_1(t) = e^{\lambda t}u \text{ and } q_2(t) = e^{\lambda t}v$$

into these equations, and let $t \to \infty$, then we have

$$\begin{aligned}\lambda u &= K_1\left[\gamma_1 (1+\lambda T_1)^{-1} v - u\right], \\ \lambda v &= K_2\left[\gamma_2 (1+\lambda T_2)^{-1} u - v\right],\end{aligned}$$

that is,

$$\begin{aligned}(\lambda + K_1)(1+\lambda T_1)u - K_1\gamma_1 v &= 0, \\ -K_2\gamma_2 u + (\lambda + K_2)(1+\lambda T_2)v &= 0.\end{aligned}$$

Therefore, the characteristic polynomial becomes

$$\begin{aligned}\varphi(\lambda) &= \det\begin{pmatrix}(\lambda + K_1)(1+\lambda T_1) & -K_1\gamma_1 \\ -K_2\gamma_2 & (\lambda + K_2)(1+\lambda T_2)\end{pmatrix} \\ &= (\lambda + K_1)(1+\lambda T_1)(\lambda + K_2)(1+\lambda T_2) - K_1K_2\gamma_1\gamma_2 = 0.\end{aligned} \tag{3.101}$$

This is an equation of degree four, which can be written as a polynomial equation,

$$a_0\lambda^4 + a_1\lambda^3 + a_2\lambda^2 + a_3\lambda + a_4 = 0$$

where

$$\begin{aligned}
a_0 &= T_1 T_2 > 0,\\
a_1 &= T_1 + T_2 + (K_1 + K_2)\, T_1 T_2 > 0,\\
a_2 &= 1 + K_1 K_2 T_1 T_2 + (K_1 + K_2)\,(T_1 + T_2) > 0,\\
a_3 &= K_1 + K_2 + K_1 K_2\,(T_1 + T_2) > 0,\\
a_4 &= K_1 K_2\,(1 - \gamma_1\gamma_2)\,.
\end{aligned}$$

We proved earlier that $\gamma_1\gamma_2 < 1/4$, so a_4 is also positive. The principal minors of the Hurwitz matrix are as follows:

$$\boldsymbol{H}_1 = a_1,\ \det(\boldsymbol{H}_2) = a_1 a_2 - a_0 a_3 > 0,\ \det(\boldsymbol{H}_4) = a_4 \det(\boldsymbol{H}_3)\,.$$

Since all coefficients are positive, all roots are negative or have negative real parts if and only if $\det(\boldsymbol{H}_3) > 0$.

Proposition 3.20 *The equilibrium with model* (3.99) *is locally asymptotically stable if and only if*

$$\gamma_1\gamma_2 > -\frac{(K_1 + K_2)\,(T_1 + T_2)\,(1 + K_1 T_1)\,(1 + K_2 T_1)\,(1 + K_1 T_2)\,(1 + K_2 T_2)}{K_1 K_2\left[T_1 + T_2 + T_1 T_2\,(K_1 + K_2)^2\right]}. \tag{3.102}$$

This condition can be easily derived by determining the form of $\det(\boldsymbol{H}_3)$ and requiring its positivity.

Consider next again the hyperbolic duopoly when firm 1 is output-adjusting and firm 2 is price-adjusting. Then in dynamic equation (3.81), $p_2(t - \tau_1)$ and $q_1(t - \tau_2)$ are replaced by

$$\int_0^t W(t - s, T_1)p_2(s)ds \text{ and } \int_0^t W(t - s, T_2)q_1(s)ds$$

where the same exponential weighting function is selected as in the previous case. The linearized equations are similar to (3.100),

$$\begin{aligned}
\dot{q}_1(t) &= K_1\left[\gamma_1 \textstyle\int_0^t W(t - s, T_1)p_2(s)ds - q_1(t)\right],\\
\dot{p}_2(t) &= K_2\left[\gamma_2 \textstyle\int_0^t W(t - s, T_2)q_1(s)ds - p_2(t)\right].
\end{aligned} \tag{3.103}$$

where γ_1 and γ_2 are now the derivatives of the best response functions (3.76) and (3.77). The characteristic equation is identical with (3.101) with the only difference that the derivatives γ_1 and γ_2 are different and $\gamma_1\gamma_2 < 1/4$ as shown earlier. So Proposition 3.20 remains the same with the only difference that γ_1 and γ_2 are now different.

In the case of Bertrand duopolies, $\gamma_1\gamma_2 = 1/4$, so (3.102) is satisfied, since the right hand side is negative. Therefore, the equilibrium in this case is always locally asymptotically stable.

3.4 Supplementary Notes and Discussions

All results developed for oligopolies can be applied to duopolies by assuming only two firms. Therefore, the developments presented in Okuguchi (1976), Okuguchi and Szidarovszky (1999) and those discussed in Bischi et al. (2010) can be specialized in the two-firm case. Especially the method of critical curves for global stability of two-dimensional systems is very useful. Its details are discussed in Bischi et al. (2010). Since there is no application of this method for delay system, it is not presented in this book.

Section 3.1 is based on two papers, Matsumoto and Szidarovszky (2007, 2010) with several new added material. Section 3.2 uses material from Matsumoto et al. (2011), Matsumoto and Szidarovszky (2015c, d), where the method of Lin and Wang (2012) is applied in the three-delay case. Its details are given in Appendix A. The material of Sect. 3.3 is mainly based on Matsumoto and Szidarovszky (2007) and Matsumoto et al. (2011) with some added new material.

Discrete time duopoly with heterogeneous expectations is examined in Agiza and Elsadany (2003) and in Elsadany and Matouk (2014), where one firm makes output decisions based on delayed information on the competitor's output but the other firm has no delay. Cournot and Bertrand duopolies are compared in Dixit (1979) and Singh and Vives (1984). Matsumoto and Onozaki (2005) and Yousefi and Szidarovszky (2005) considered mixed duopolies with nonlinear demands and showed the birth of complicated fluctuations.

Chapter 4
Dynamic Oligopolies

Oligopoly models describe the behavior of n firms who produce related goods or offer related services to a homogeneous market. They are generalizations of duopoly models with arbitrary number of firms.

One of the most frequently analyzed models can be obtained as follows. Consider a representative consumer, and assume that his/her utility function is given as

$$u = \sum_{k=1}^{n} \alpha_k q_k - \frac{1}{2}\left(\sum_{k=1}^{n} q_k^2 + 2\gamma \sum_{k=1}^{n-1}\sum_{i>k} q_k q_i\right) \tag{4.1}$$

where q_k is the output of firm k, α_k measures the quality of product k and γ indicates the substitutability between the goods with $-1 \le \gamma \le 1$. If $\gamma > 0$, then the goods are substitutes, if $\gamma < 0$, then they are complements and if $\gamma = 0$, then the goods are independent. If p_k denotes the unit price of good k, then the representative consumer maximizes the difference of the utility and total cost:

$$\sum_{k=1}^{n} \alpha_k q_k - \frac{1}{2}\left(\sum_{k=1}^{n} q_k^2 + 2\gamma \sum_{k=1}^{n-1}\sum_{i>k} q_k q_i\right) - \sum_{k=1}^{n} p_k q_k. \tag{4.2}$$

The first-order conditions imply that the price function of product k is the following:

$$p_k = \alpha_k - q_k - \gamma \sum_{i \ne k}^{n} q_i \quad (1 \le k \le n). \tag{4.3}$$

Therefore, the profit function of firm k is given as

A. Matsumoto and F. Szidarovszky, *Dynamic Oligopolies with Time Delays*,
https://doi.org/10.1007/978-981-13-1786-6_4

$$\pi_k(q_1, \ldots, q_n) = q_k \left(\alpha_k - q_k - \gamma \sum_{i \neq k}^{n} q_i - c_k \right) \tag{4.4}$$

where linear cost function is assumed with zero fixed cost and marginal cost $c_k > 0$. If $\gamma = 0$, then each firm is a monopoly being independent from the others, so we may assume that $\gamma \neq 0$. This model is called quantity-adjusting (Cournot) oligopoly with product differentiation. In the special case, when $\alpha_1 = \cdots = \alpha_n = \alpha$ and $\gamma = 1$, the model is without product differentiation. In order to exclude negative profit, we assume that $\alpha_k > c_k$.

If the firms select prices as decision variables, then the model is called price-adjusting (Bertrand) oligopoly. The associated demands of the differentiated products can be obtained by solving equations (4.3) for unknowns $q_1, \ldots, q_n$ with given prices $p_1, \ldots, p_n$. Assuming that

$$\gamma \neq 1 \text{ and } \gamma \neq -\frac{1}{n-1},$$

we have

$$q_k = \frac{[1 + (n-2)\gamma](\alpha_k - p_k) - \gamma \sum_{i \neq k}^{n} (\alpha_i - p_i)}{(1-\gamma)[1 + (n-1)\gamma]}. \tag{4.5}$$

This relation can be proved as follows. Introduce vectors $\boldsymbol{\alpha} = (\alpha_k)$, $\boldsymbol{p} = (p_k)$ and $\boldsymbol{q} = (q_k)$, and let matrix $\boldsymbol{B} = (b_{ij})$ be defined as $b_{ii} = 1$ and $b_{ij} = \gamma$ for $j \neq i$. Then Eq. (4.3) can be rewritten as

$$\boldsymbol{p} = \boldsymbol{\alpha} - \boldsymbol{B}\boldsymbol{q} \tag{4.6}$$

so

$$\boldsymbol{q} = \boldsymbol{B}^{-1}(\boldsymbol{\alpha} - \boldsymbol{p})$$

where the diagonal elements of $\boldsymbol{B}^{-1}$ are

$$\frac{1 + (n-2)\gamma}{(1-\gamma)[1 + (n-1)\gamma]}$$

and the off-diagonal elements are all equal,

$$-\frac{\gamma}{(1-\gamma)[1 + (n-1)\gamma]}$$

which can be verified by checking that $\boldsymbol{B}^{-1}\boldsymbol{B}$ is the identity matrix.

In the resulting n-firm Bertrand oligopoly, the profit of firm k is the following:

$$\bar{\pi}_k(p_1, \ldots, p_n) = \frac{[1 + (n-2)\gamma](\alpha_k - p_k) - \gamma \sum_{i \neq k}^{n} (\alpha_i - p_i)}{(1-\gamma)[1 + (n-1)\gamma]}(p_k - c_k). \tag{4.7}$$

Simple differentiation shows that from (4.4)

$$\frac{\partial \pi_k(q_1, \ldots, q_n)}{\partial q_k} = \alpha_k - 2q_k - \gamma \sum_{i \neq k}^{n} q_i - c_k$$

and

$$\frac{\partial^2 \pi_k(q_1, \ldots, q_n)}{\partial q_k^2} = -2 < 0.$$

Similarly, from (4.7),

$$\frac{\partial \bar{\pi}_k(p_1, \ldots, p_n)}{\partial p_k} = \frac{-[1 + (n-2)\gamma](p_k - c_k) + [1 + (n-2)\gamma](\alpha_k - p_k) - \gamma \sum_{i \neq k}^{n} (\alpha_i - p_i)}{(1-\gamma)[1 + (n-1)\gamma]}$$

and

$$\frac{\partial^2 \bar{\pi}_k(p_1, \ldots, p_n)}{\partial p_k^2} = -\frac{2[1 + (n-2)\gamma]}{(1-\gamma)[1 + (n-1)\gamma]}.$$

Therefore, the profit function of the Cournot firms are strictly concave in their outputs; however, the profits of Bertrand firms are concave only if either

$$\gamma < -\frac{1}{n-2} \text{ or } \gamma > -\frac{1}{n-1}.$$

Otherwise, the first-order conditions do not imply that the profits are on their maximum levels. The case when

$$-\frac{1}{n-2} < \gamma < -\frac{1}{n-1}$$

is therefore eliminated from our consideration.

We will next introduce mixed oligopolies. Assume that firms $k \in K \subset \{1, 2, \ldots, n\}$ are quantity setting and firms $\bar{K} = \{1, 2, \ldots, n\} \backslash K$ are price setting. If $\bar{K}$ is empty, then the oligopoly is Cournot, and if K is empty, then it is Bertrand. Without losing generality, we may assume that the quantity setting firms are $1, 2, \ldots, r$ and the price-setting firms are $r+1, \ldots, n$. In order to have a mixed oligopoly, we assume that

$$1 \leq r \leq n - 1.$$

Introduce the following notation

$$\boldsymbol{q}^K = (q_k)_{k \in K} \text{ and } \boldsymbol{q}^{\bar{K}} = (q_k)_{k \in \bar{K}}$$

$$\boldsymbol{p}^K = (p_k)_{k \in K} \text{ and } \boldsymbol{p}^{\bar{K}} = (p_k)_{k \in \bar{K}}$$

$$\boldsymbol{\alpha}^K = (\alpha_k)_{k\in K} \text{ and } \boldsymbol{\alpha}^{\bar{K}} = (\alpha_k)_{k\in\bar{K}}\,.$$

Then Eq. (4.6) can be rewritten as the following block form:

$$\begin{pmatrix} \boldsymbol{p}^K \\ \boldsymbol{p}^{\bar{K}} \end{pmatrix} = \begin{pmatrix} \alpha^K \\ \alpha^{\bar{K}} \end{pmatrix} - \begin{pmatrix} \boldsymbol{B}^{KK} & \boldsymbol{B}^{K\bar{K}} \\ \boldsymbol{B}^{\bar{K}K} & \boldsymbol{B}^{\bar{K}\bar{K}} \end{pmatrix} \begin{pmatrix} \boldsymbol{q}^K \\ \boldsymbol{q}^{\bar{K}} \end{pmatrix}$$

where $\boldsymbol{B}$ is broken up to blocks accordingly. That is,

$$\boldsymbol{p}^K = \boldsymbol{\alpha}^K - \boldsymbol{B}^{KK}\boldsymbol{q}^K - \boldsymbol{B}^{K\bar{K}}\boldsymbol{q}^{\bar{K}} \tag{4.8}$$

and

$$\boldsymbol{p}^{\bar{K}} = \boldsymbol{\alpha}^{\bar{K}} - \boldsymbol{B}^{\bar{K}K}\boldsymbol{q}^K - \boldsymbol{B}^{\bar{K}\bar{K}}\boldsymbol{q}^{\bar{K}}. \tag{4.9}$$

We solve next these equations for unknowns $\boldsymbol{p}^K$ and $\boldsymbol{q}^{\bar{K}}$. From (4.9), we have

$$\boldsymbol{q}^{\bar{K}} = \left(\boldsymbol{B}^{\bar{K}\bar{K}}\right)^{-1}\left(\alpha^{\bar{K}} - \boldsymbol{B}^{\bar{K}K}\boldsymbol{q}^K - \boldsymbol{p}^{\bar{K}}\right) \tag{4.10}$$

and by substituting it into Eq. (4.8), we can obtain $\boldsymbol{p}^{\bar{K}}$ as a function of $\boldsymbol{q}^K$ and $\boldsymbol{p}^{\bar{K}}$. The expressions become complicated, therefore for the sake of simplicity, we introduce the following notation:

$$\boldsymbol{A}^K = \boldsymbol{\alpha}^K - \boldsymbol{B}^{K\bar{K}}\left(\boldsymbol{B}^{\bar{K}\bar{K}}\right)^{-1}\alpha^{\bar{K}} \text{ and } \boldsymbol{A}^{\bar{K}} = \left(\boldsymbol{B}^{\bar{K}\bar{K}}\right)^{-1}\alpha^{\bar{K}},$$

$$\beta^{KK} = \boldsymbol{B}^{KK} - \boldsymbol{B}^{K\bar{K}}\left(\boldsymbol{B}^{\bar{K}\bar{K}}\right)^{-1}\boldsymbol{B}^{\bar{K}K}, \beta^{K\bar{K}} = -\boldsymbol{B}^{K\bar{K}}\left(\boldsymbol{B}^{\bar{K}\bar{K}}\right)^{-1},$$

$$\beta^{\bar{K}K} = \left(\boldsymbol{B}^{\bar{K}\bar{K}}\right)^{-1}\boldsymbol{B}^{\bar{K}K} \text{ and } \beta^{\bar{K}\bar{K}} = \left(\boldsymbol{B}^{\bar{K}\bar{K}}\right)^{-1}.$$

Then it is easy to see that

$$\boldsymbol{p}^K = \boldsymbol{A}^K - \beta^{KK}\boldsymbol{q}^K - \beta^{K\bar{K}}\boldsymbol{p}^{\bar{K}} \tag{4.11}$$

and

$$\boldsymbol{q}^{\bar{K}} = \boldsymbol{A}^{\bar{K}} - \beta^{\bar{K}K}\boldsymbol{q}^K - \beta^{\bar{K}\bar{K}}\boldsymbol{p}^{\bar{K}}. \tag{4.12}$$

Observe that the components of $\boldsymbol{p}^K$ and $\boldsymbol{q}^{\bar{K}}$ are linear functions of the components of $\boldsymbol{q}^K$ and $\boldsymbol{p}^{\bar{K}}$. The profit functions of firm $k \in K$ are as follows:

$$\bar{\pi}'_k\left(\boldsymbol{q}^K, \boldsymbol{p}^{\bar{K}}\right) = \left(A_k - \sum_{j=1}^{r} \beta_{kj}^{KK} q_j - \sum_{j=r+1}^{n} \beta_{kj}^{K\bar{K}} p_j - c_k\right) q_k \tag{4.13}$$

and the profits of firms $k \in \bar{K}$ are similarly

$$\bar{\pi}'_k\left(\boldsymbol{q}^K, \boldsymbol{p}^{\bar{K}}\right) = \left(A_k - \sum_{j=1}^{r} \beta_{kj}^{\bar{K}K} q_j - \sum_{j=r+1}^{n} \beta_{kj}^{\bar{K}\bar{K}} p_j\right) (p_k - c_k). \tag{4.14}$$

Notice that the profit functions, (4.7), (4.13) and (4.14) have the same linear structure as (4.4), so their dynamic analysis can be done similarly. In addition,

$$\frac{\partial \bar{\pi}'_k}{\partial q_k} = A_k - \sum_{j=1, j\neq k}^{r} \beta_{kj}^{KK} q_j - 2\beta_{kk}^{KK} q_k - \sum_{j=r+1}^{n} \beta_{kj}^{K\bar{K}} p_j - c_k \quad (1 \leq k \leq r)$$

and

$$\frac{\partial^2 \bar{\pi}'_k}{\partial q_k^2} = -2\beta_{kk}^{KK},$$

so $\bar{\pi}'_k$ is concave in q_k if $\beta_{kk}^{KK} > 0$. Similarly,

$$\frac{\partial \bar{\pi}'_k}{\partial p_k} = A_k - \sum_{j=1}^{r} \beta_{kj}^{\bar{K}K} q_j - \sum_{j=r+1}^{n} \beta_{kj}^{\bar{K}\bar{K}} p_j - (p_k - c_k)\beta_{kk}^{\bar{K}\bar{K}} \quad (r+1 \leq k \leq n)$$

and

$$\frac{\partial^2 \bar{\pi}'_k}{\partial p_k^2} = -2\beta_{kk}^{\bar{K}\bar{K}}$$

which is negative if $\beta_{kk}^{\bar{K}\bar{K}} > 0$, implying the concavity of $\bar{\pi}'_k$ in p_k.

Assume next that the firms produce homogenous goods. Let $Q = \sum_{k=1}^{n} q_k$ be the output of the industry. Assume in addition that the representative consumer has an isoelastic utility functions

$$u(Q) = \begin{cases} \dfrac{Q^{1-\eta} - 1}{1-\eta} & \text{if } \eta \neq 1, \\[2ex] \ln(Q) & \text{if } \eta = 1. \end{cases} \tag{4.15}$$

Notice that by the use of the L'Hospital's rule,

$$\lim_{\eta\to 1} \frac{Q^{1-\eta} - 1}{1-\eta} = \lim_{\eta\to 1} \frac{Q^{1-\eta}\ln(Q)(-1)}{-1} = \ln Q.$$

If p is the common price of the goods, then the consumer maximizes

$$\frac{Q^{1-\eta}-1}{1-\eta} - pQ \text{ if } \eta \neq 1$$

and

$$\ln(Q) - pQ \text{ if } \eta = 1$$

implying that

$$p = \frac{1}{Q^{\eta}} \tag{4.16}$$

in both cases leading to a hyperbolic price function. If the products are differentiated, then let Θ_{ij} denote the substitutability of product i by product j, then with $\eta = 1$ the price of product k becomes

$$p_k = \frac{1}{q_k + \sum_{j \neq k} \Theta_{kj} q_j}. \tag{4.17}$$

In the special case of $\Theta_{kj} = 1$ for all k and j, the oligopoly is without product differentiation.

For each firm, let

$$Q_k = \sum_{j \neq k}^{n} \Theta_{kj} q_j,$$

then the profit of firm k is given as

$$\pi_k(q_1, \ldots, q_n) = \frac{q_k}{q_k + Q_k} - c_k q_k \tag{4.18}$$

where linear cost function is assumed with zero fixed cost.

Assume next that the firms select prices as decision variables. Then the corresponding demands can be derived from Eq. (4.17) which is a system of linear equations for the unknown variables $q_1, \ldots, q_n$:

$$q_k + \sum_{j \neq k} \Theta_{kj} q_j = \frac{1}{p_k}. \tag{4.19}$$

With coefficient matrix

$$\boldsymbol{B} = \begin{pmatrix} 1 & \Theta_{12} & \cdots & \Theta_{1n} \\ \Theta_{21} & 1 & \cdots & \Theta_{2n} \\ \cdot & \cdot & \cdots & \cdot \\ \Theta_{n1} & \Theta_{n2} & \cdots & 1 \end{pmatrix},$$

the solution of (4.19) is obtained as

$$q = B^{-1}\bar{p}$$

where

$$q = (q_k) \text{ and } \bar{p} = \left(\frac{1}{p_k}\right).$$

Let b_{ij}^{-1} denote the (i, j) element of B^{-1}, then

$$q_k = \sum_{j=1}^{n} b_{kj}^{-1}\frac{1}{p_j} \tag{4.20}$$

and therefore the profit of firm k is as follows:

$$\bar{\pi}_k(p_1, \ldots, p_n) = \left(\sum_{j=1}^{n} b_{kj}^{-1}\frac{1}{p_j}\right)(p_k - c_k). \tag{4.21}$$

Simple differentiation shows that

$$\frac{\partial \pi_k(q_1, \ldots, q_n)}{\partial q_k} = \frac{Q_k}{(q_k + Q_k)^2} - c_k$$

and

$$\frac{\partial^2 \pi_k(q_1, \ldots, q_n)}{\partial q_k^2} = -\frac{2Q_k}{(q_k + Q_k)^3} < 0$$

and

$$\frac{\partial \bar{\pi}_k(p_1, \ldots, p_n)}{\partial p_k} = \left(\sum_{j=1}^{n} b_{kj}^{-1}\frac{1}{p_j}\right) + (p_k - c_k)\left(-\frac{b_{kk}^{-1}}{p_k^2}\right)$$

$$= \sum_{j \neq k}\frac{b_{kj}^{-1}}{p_j} + \frac{c_k b_{kk}^{-1}}{p_k^2}$$

and

$$\frac{\partial^2 \bar{\pi}_k(p_1, \ldots, p_n)}{\partial p_k^2} = -\frac{2c_k b_{kk}^{-1}}{p_k^3}.$$

The profits of the firms in the Cournot case are strictly concave in the outputs; however, in the case of Bertrand oligopolies, $\bar{\pi}_k$ is concave in p_k if $b_{kk}^{-1} > 0$. This is the case, for example, if the Θ_{kj} values are sufficiently small, so B^{-1} is close to the identity matrix.

If the hyperbolic oligopoly is without product differentiation, then there is only one price function, $p = 1/Q$, when Q is the industry output, $Q_k = \sum_{j \neq k} q_j$ is the output of the rest of the industry and the profit of firm k in the Cournot competition is given by Eq. (4.18). Since there is only one price, no Bertrand competition can be defined.

We will next introduce mixed oligopoly similarly to the linear case discussed earlier. Assume that firms $k \in K \subset \{1, 2, \ldots, n\}$ are quantity setting and firms from $\bar{K} = \{1, 2, \ldots, n\} \backslash K$ are price setting, where both K and $\bar{K}$ are nonempty. We can also assume that firms $1, 2, \ldots, r$ are quantity setting and firms $r+1, \ldots, n$ are price setting with $1 \le r \le n-1$. Introduce now the notations:

$$\boldsymbol{q}^K = (q_k)_{k \in K} \text{ and } \boldsymbol{q}^{\bar{K}} = (q_k)_{k \in \bar{K}} ,$$

$$\bar{\boldsymbol{p}}^K = \left(\frac{1}{p_k}\right)_{k \in K} \text{ and } \bar{\boldsymbol{p}}^{\bar{K}} = \left(\frac{1}{p_i}\right)_{i \in \bar{K}} .$$

Then Eq. (3.19) can be written as

$$\begin{pmatrix} \boldsymbol{B}^{KK} & \boldsymbol{B}^{K\bar{K}} \\ \boldsymbol{B}^{\bar{K}K} & \boldsymbol{B}^{\bar{K}\bar{K}} \end{pmatrix} \begin{pmatrix} \boldsymbol{q}^K \\ \boldsymbol{q}^{\bar{K}} \end{pmatrix} = \begin{pmatrix} \bar{\boldsymbol{p}}^K \\ \bar{\boldsymbol{p}}^{\bar{K}} \end{pmatrix}$$

where matrix $\boldsymbol{B}$ is broken up to blocks according to the sizes of vectors $\boldsymbol{q}^K$ and $\boldsymbol{q}^{\bar{K}}$. So

$$\boldsymbol{B}^{KK}\boldsymbol{q}^K + \boldsymbol{B}^{K\bar{K}}\boldsymbol{q}^{\bar{K}} = \bar{\boldsymbol{p}}^K \tag{4.22}$$

and

$$\boldsymbol{B}^{\bar{K}K}\boldsymbol{q}^K + \boldsymbol{B}^{\bar{K}\bar{K}}\boldsymbol{q}^{\bar{K}} = \bar{\boldsymbol{p}}^{\bar{K}}. \tag{4.23}$$

From (4.23) we have

$$\boldsymbol{q}^{\bar{K}} = \left(\boldsymbol{B}^{\bar{K}\bar{K}}\right)^{-1} \left(\bar{\boldsymbol{p}}^{\bar{K}} - \boldsymbol{B}^{\bar{K}K}\boldsymbol{q}^K\right)$$

and then from (4.22),

$$\begin{aligned} \bar{\boldsymbol{p}}^K &= \boldsymbol{B}^{KK}\boldsymbol{q}^K + \boldsymbol{B}^{K\bar{K}} \left(\boldsymbol{B}^{\bar{K}\bar{K}}\right)^{-1} \left(\bar{\boldsymbol{p}}^{\bar{K}} - \boldsymbol{B}^{\bar{K}K}\boldsymbol{q}^K\right) \\ &= \boldsymbol{B}^{K\bar{K}} \left(\boldsymbol{B}^{\bar{K}\bar{K}}\right)^{-1} \bar{\boldsymbol{p}}^{\bar{K}} + \left(\boldsymbol{B}^{KK} - \boldsymbol{B}^{K\bar{K}} \left(\boldsymbol{B}^{\bar{K}\bar{K}}\right)^{-1} \boldsymbol{B}^{\bar{K}K}\right) \boldsymbol{q}^K. \end{aligned}$$

Introduce now the additional notation

$$\beta^{KK} = \boldsymbol{B}^{KK} - \boldsymbol{B}^{K\bar{K}} \left(\boldsymbol{B}^{\bar{K}\bar{K}}\right)^{-1} \boldsymbol{B}^{\bar{K}K}, \beta^{K\bar{K}} = \boldsymbol{B}^{K\bar{K}} \left(\boldsymbol{B}^{\bar{K}\bar{K}}\right)^{-1},$$

$$\beta^{\bar{K}K} = -\left(\boldsymbol{B}^{\bar{K}\bar{K}}\right)^{-1} \boldsymbol{B}^{\bar{K}K} \text{ and } \beta^{\bar{K}\bar{K}} = \left(\boldsymbol{B}^{\bar{K}\bar{K}}\right)^{-1},$$

then we can easily see that

$$\bar{\boldsymbol{p}}^K = \boldsymbol{\beta}^{KK}\boldsymbol{q}^K + \boldsymbol{\beta}^{K\bar{K}}\bar{\boldsymbol{p}}^{\bar{K}}$$

and

$$\boldsymbol{q}^{\bar{K}} = \boldsymbol{\beta}^{\bar{K}K}\boldsymbol{q}^K + \boldsymbol{\beta}^{\bar{K}\bar{K}}\bar{\boldsymbol{p}}^{\bar{K}}.$$

Therefore, the profits of firms $k \in K$ are given as

$$\bar{\pi}'_k(\boldsymbol{q}^K, \bar{\boldsymbol{p}}^{\bar{K}}) = \left(\sum_{j=1}^{r} \beta_{kj}^{KK} q_j + \sum_{j=r+1}^{n} \beta_{kj}^{K\bar{K}} \frac{1}{p_j} - c_k\right) q_k \tag{4.24}$$

and those of firms $k \in \bar{K}$,

$$\bar{\pi}'_k(\boldsymbol{q}^K, \bar{\boldsymbol{p}}^{\bar{K}}) = \left(\sum_{j=1}^{r} \beta_{kj}^{\bar{K}K} q_j + \sum_{j=r+1}^{n} \beta_{kj}^{\bar{K}\bar{K}} \frac{1}{p_j}\right) (p_k - c_k)\,. \tag{4.25}$$

Observe that both profit functions are linear in variables q_j $(j = 1, 2, \ldots, r)$ and $1/p_j$ $(j = r+1, \ldots, n)$.

Notice that from (4.24) for $k = 1, 2, \ldots, r$,

$$\frac{\partial \bar{\pi}'_k}{\partial q_k} = \sum_{j=1}^{r} \beta_{kj}^{KK} q_j + \sum_{j=r+1}^{n} \beta_{kj}^{K\bar{K}} \frac{1}{p_j} - c_k + \beta_{kk}^{KK} q_k$$

so

$$\frac{\partial^2 \bar{\pi}'_k}{\partial q_k^2} = 2\beta_{kk}^{KK}$$

which is negative, or $\bar{\pi}'_k$ is concave in q_k, if $\beta_{kk}^{KK} < 0$. From (4.25), we have similarly for $k = r+1, r+2, \ldots, n$,

$$\frac{\partial \bar{\pi}'_k}{\partial p_k} = \sum_{j=1}^{r} \beta_{kj}^{\bar{K}K} q_j + \sum_{j=r+1}^{n} \beta_{kj}^{\bar{K}\bar{K}} \frac{1}{p_j} + (p_k - c_k)\left(-\frac{\beta_{kk}^{\bar{K}\bar{K}}}{p_k^2}\right)$$

therefore

$$\frac{\partial^2 \bar{\pi}'_k}{\partial p_k^2} = -\frac{2c_k \beta_{kk}^{\bar{K}\bar{K}}}{p_k^3}$$

which is negative, or $\bar{\pi}'_k$ is concave in p_k, if $\beta_{kk}^{\bar{K}\bar{K}} > 0$.

In the above derivation, we assumed that matrix $\boldsymbol{B}$ is invertible. In the case of models without product differentiation $\Theta_{kj} = 1$ for all k and j, so all elements of

$\boldsymbol{B}$ are equal to 1, therefore $\boldsymbol{B}$ is not invertible. This is logical, since in an oligopoly without product differentiation, no price-setting firms can be present.

4.1 Discrete Dynamics

Consider first an n-firm linear Cournot oligopoly with product differentiation. The profit of firm k is given in Eq. (4.4), so the marginal profit equals

$$\frac{\partial \pi_k}{\partial q_k} = \alpha_k - 2q_k - \gamma \sum_{i \neq k} q_i - c_k$$

as it was derived earlier. So by assuming interior optimum, the best response of firm k becomes

$$q_k = R_k\left(Q_k\right) = \frac{\alpha_k - \gamma Q_k - c_k}{2} \tag{4.26}$$

where $Q_k = \sum_{i \neq k} q_i$ is the output of the rest of the industry from the point of view of firm k. Notice that $R_k\left(Q_k\right)$ can be rewritten as

$$q_k = R_k\left(Q_k\right) = \frac{\alpha_k - c_k - \gamma Q + \gamma q_k}{2}$$

where $Q = \sum_{k=1}^{n} q_k$ is the industry output. Therefore,

$$q_k = \frac{\alpha_k - c_k - \gamma Q}{2 - \gamma}. \tag{4.27}$$

By adding this equation for $k = 1, 2, \ldots, n$, we have

$$Q = \frac{1}{2 - \gamma}\left[\sum_{k=1}^{n} \left(\alpha_k - c_k\right) - n\gamma Q\right]$$

implying that the equilibrium industry output is equal to

$$Q = \frac{\sum_{k=1}^{n} \left(\alpha_k - c_k\right)}{2 + (n-1)\gamma}. \tag{4.28}$$

We assume that

$$\gamma > -\frac{2}{n-1}$$

to have positive industry output. From (4.27),

$$q_k = \frac{\alpha_k - c_k}{2-\gamma} - \frac{\gamma}{(2-\gamma)\left[2+(n-1)\gamma\right]} \sum_{i=1}^{n} (\alpha_i - c_i) \tag{4.29}$$

and the associated market price

$$p_k = \frac{\alpha_k + c_k - \gamma c_k}{2-\gamma} - \frac{\gamma}{(2-\gamma)\left[2+(n-1)\gamma\right]} \sum_{i=1}^{n} (\alpha_i - c_i)\,. \tag{4.30}$$

The equilibrium output of firm k is positive, if

$$\alpha_k - c_k > \frac{\gamma n}{2+(n-1)\gamma}\left[\frac{1}{n}\sum_{i=1}^{n}(\alpha_i - c_i)\right].$$

Notice that $\alpha_k - c_k$ is the difference of the quality indicator (or maximum price) and unit production cost, so larger difference shows that firm k is better qualified, since with same cost it produces better product. The bracketed factor in the right hand side is the average qualification of all firms.

We assumed that $\gamma > -2/(n-1)$ and notice that in case of $-2/(n-1) < \gamma < 0$, the first factor ranges between $-\infty$ and 0, and if $0 \le \gamma \le 1$, then its value is between 0 and $n/(n+1)$.

In developing the dynamic process, assume that the firms use best response dynamics and think that there is no output change of the competitors between time periods t and $t+1$. Then the dynamic equations become

$$q_k(t+1) = R_k\left(Q_k(t)\right) = \frac{\alpha_k - c_k - \gamma Q_k(t)}{2} \quad (k = 1, 2, \ldots, n) \tag{4.31}$$

with Jacobian matrix

$$\boldsymbol{J}_1 = \begin{pmatrix} 0 & -\frac{\gamma}{2} & \ldots & -\frac{\gamma}{2} \\ -\frac{\gamma}{2} & 0 & \ldots & -\frac{\gamma}{2} \\ \cdot & \cdot & \ldots & \cdot \\ -\frac{\gamma}{2} & -\frac{\gamma}{2} & \ldots & 0 \end{pmatrix}.$$

If partial adjustments toward best response is assumed, then (4.31) is modified as

$$q_k(t+1) = (1 - K_k)\, q_k(t) + K_k R_k \left(\sum_{i \ne k} q_i(t)\right) \quad (k = 1, 2, \ldots, n) \tag{4.32}$$

where $0 < K_k \le 1$ is assumed. The Jacobian of this system is

$$J = \begin{pmatrix} 1-K_1 & -\frac{K_1\gamma}{2} & \dots & -\frac{K_1\gamma}{2} \\ -\frac{K_2\gamma}{2} & 1-K_2 & \dots & -\frac{K_2\gamma}{2} \\ \cdot & \cdot & \dots & \cdot \\ -\frac{K_n\gamma}{2} & -\frac{K_n\gamma}{2} & \dots & 1-K_n \end{pmatrix}.$$

The eigenvalues are inside the unit circle if a norm of matrix J is less than one. The row norm is less than one, if for all k,

$$\frac{K_k}{2}(n-1)\,|\gamma| < K_k$$

that is,

$$|\gamma| < \frac{2}{n-1} \text{ or } -\frac{2}{n-1} < \gamma < \frac{2}{n-1}.$$

The column norm is below one, if for all k,

$$\frac{1}{2}\left(\sum_{i \neq k} K_i\right)|\gamma| < K_k.$$

These are only simple sufficient conditions; however, sufficient and necessary conditions will be derived next.

If $K_1 = \cdots = K_n = 1$, then matrix J is simplified to J_1. Notice that

$$J = D + ab^T \tag{4.33}$$

where

$$D = diag\left(1 - K_1\left(1 - \frac{\gamma}{2}\right), \dots, 1 - K_n\left(1 - \frac{\gamma}{2}\right)\right)$$

$$a = \left(-\frac{K_1\gamma}{2}, \dots, -\frac{K_n\gamma}{2}\right)^T$$

and

$$b^T = (1, \dots, 1).$$

Then by using the well-known relation (see, for example, Bischi et al. 2010) that with n-element vectors u and v, and the $n \times n$ identical matrix I,

$$\det\left(I + uv^T\right) = 1 + u^T v, \tag{4.34}$$

the characteristic polynomial of matrix J is as follows:

$$\varphi(\lambda) = \det\left(\boldsymbol{D} + \boldsymbol{a}\boldsymbol{b}^T - \lambda \boldsymbol{I}\right)$$

$$= \det\left(\boldsymbol{D} - \lambda \boldsymbol{I}\right) \det\left(\boldsymbol{I} + \left(\boldsymbol{D} - \boldsymbol{\lambda} \boldsymbol{I}\right)^{-1} \boldsymbol{a}\boldsymbol{b}^T\right)$$

$$= \prod_{k=1}^{n} \left(1 - K_k\left(1 - \frac{\gamma}{2}\right) - \lambda\right) \left[1 + \sum_{k=1}^{n} \frac{-\frac{K_k\gamma}{2}}{1 - K_k\left(1 - \frac{\gamma}{2}\right) - \lambda}\right].$$

Assume that the firms are numbered such that the different K_k values satisfy

$$K_1 > K_2 > \cdots > K_S$$

with multiplicities $m_1, m_2, \ldots, m_S$. By adding the firms with identical denominators in the bracketed expression, the characteristic equation becomes

$$\prod_{j=1}^{S} \left(1 - K_j\left(1 - \frac{\gamma}{2}\right) - \lambda\right)^{m_j} \left[1 + \sum_{j=1}^{S} \frac{-\frac{T_j\gamma}{2}}{1 - K_j\left(1 - \frac{\gamma}{2}\right) - \lambda}\right] = 0 \qquad (4.35)$$

where T_j is the sum of the identical K_k values. If $m_j \geq 2$, then $1 - K_j\left(1 - \frac{\gamma}{2}\right)$ is an eigenvalue, and if $m_j = 1$, then $1 - K_j\left(1 - \frac{\gamma}{2}\right)$ is not an eigenvalue. All values $1 - K_j\left(1 - \frac{\gamma}{2}\right)$ are between -1 and $+1$, since $0 < K_j \leq 1$ and $-1 \leq \gamma \leq 1$ as assumed before. The other eigenvalues are the solutions of equation

$$g(\lambda) = 1 + \sum_{j=1}^{S} \frac{-\frac{T_j\gamma}{2}}{1 - K_j\left(1 - \frac{\gamma}{2}\right) - \lambda} = 0. \qquad (4.36)$$

The poles of $g(\lambda)$ are the $1 - K_j\left(1 - \frac{\gamma}{2}\right)$ values which are between -1 and $+1$. We consider now two cases.

If $\gamma > 0$, then $g'(\lambda) < 0$, and if λ converges to any pole from the left, then $g(\lambda)$ tends to $-\infty$, and if λ converges to any pole from the right, then $g(\lambda) \to +\infty$. The graph of $g(\lambda)$ is shown in Fig. 4.1. Equation (4.36) is equivalent to a polynomial equation of degree S. There are $S - 1$ solutions between the poles, and the last solution is below the smallest pole. Therefore, all solutions are between -1 and $+1$ if and only if $g(-1) > 0$, which can be written as

$$\sum_{k=1}^{n} \frac{K_k\gamma}{4 - K_k(2 - \gamma)} < 1. \qquad (4.37)$$

If $\gamma < 0$, then $g'(\lambda) > 0$ and if λ converges to any pole from the left, then $g(\lambda) \to +\infty$ and if it converges to a pole from the right, then $g(\lambda)$ tends to $-\infty$. The graph of $g(\lambda)$ is shown in Fig. 4.2. We have $S-1$ solutions between the poles, so they are between -1 and $+1$, the last solution is above the largest pole, so all solutions are between -1 and $+1$ if and only if $g(1) > 0$. This condition can be rewritten as

$$1 + \sum_{k=1}^{n} \frac{\gamma}{2-\gamma} = 1 + \frac{n\gamma}{2-\gamma} > 0,$$

or

$$\gamma > -\frac{2}{n-1}. \tag{4.38}$$

Proposition 4.1 *The equilibrium of the linear oligopoly with model* (4.32) *is asymptotically stable if and only if*

(*a*) (4.37) *holds for* $\gamma > 0$

(*b*) (4.38) *holds for* $\gamma < 0$.

Notice that condition (4.38) was assumed to guarantee positive equilibrium industry output. If $K_1 = \cdots = K_n = 1$, then system (4.31) is obtained as a special case, in which (4.37) is specialized as $\gamma < 2/(n-1)$. So considering the two cases $\gamma > 0$ and $\gamma < 0$, we have the stability condition:

$$-\frac{2}{n-1} < \gamma < \frac{2}{n-1}.$$

Assume next that the firms have a unit period delay in obtaining information about the output of the rest of the industry. Then system (4.32) is modified as

$$q_k(t+1) = (1-K_k)\, q_k(t) + K_k R_k \left(\sum_{i \neq k} q_i(t-1) \right) \quad (k = 1, 2, \ldots, n). \tag{4.39}$$

By introducing the new variables,

$$z_k(t) = q_k(t-1),$$

a $2n$-dimensional system is obtained without delays,

$$\begin{aligned} q_k(t+1) &= (1-K_k)\, q_k(t) + K_k \frac{\alpha_k - c_k - \gamma \sum_{i \neq k} z_i(t)}{2}, \\ z_k(t+1) &= q_k(t) \quad (k = 1, 2, \ldots, n) \end{aligned} \tag{4.40}$$

with coefficient matrix

$$J = \begin{pmatrix} 1-K_1 & 0 & \dots & 0 & 0 & -\frac{K_1\gamma}{2} & \dots & -\frac{K_1\gamma}{2} \\ 0 & 1-K_2 & \dots & 0 & -\frac{K_2\gamma}{2} & 0 & \dots & -\frac{K_2\gamma}{2} \\ \cdot & \cdot & \dots & \cdot & \cdot & \cdot & \dots & \cdot \\ 0 & 0 & \dots & 1-K_n & -\frac{K_n\gamma}{2} & -\frac{K_n\gamma}{2} & \dots & 0 \\ 1 & 0 & \dots & 0 & 0 & 0 & \dots & 0 \\ 0 & 1 & \dots & 0 & 0 & 0 & \dots & 0 \\ \cdot & \cdot & \dots & \cdot & \cdot & \cdot & \dots & \cdot \\ 0 & 0 & \dots & 1 & 0 & 0 & \dots & 0 \end{pmatrix}.$$

The eigenvalue equations of this matrix can be written as follows: if the first n components of the eigenvector are denoted by u_k and the last n components by v_k, then

$$(1-K_k)\,u_k - \sum\nolimits_{i\neq k} \frac{K_k\gamma}{2} v_i = \lambda u_k$$

$$u_k = \lambda v_k$$

and by substituting the second equation into the first, we have

$$\left[\lambda^2 - \lambda\,(1-K_k)\right] v_k + \sum_{i\neq k} \frac{K_k\gamma}{2} v_i = 0 \; (k = 1, 2, \dots, n).$$

By introducing the notation

$$P_k(\lambda) = \lambda^2 - \lambda\,(1-K_k) - \frac{K_k\gamma}{2},$$

$$\boldsymbol{a} = \left(\frac{K_1\gamma}{2}, \dots, \frac{K_n\gamma}{2}\right)^T$$

and

$$\boldsymbol{b}^T = (1, 1, \dots, 1),$$

the determinant of the coefficient matrix can be written in the following form similarly to the previous case based on relation (4.34):

$$\det\left(\boldsymbol{D} + \boldsymbol{a}\boldsymbol{b}^T\right) = \det(\boldsymbol{D}) \det\left(\boldsymbol{I} + \boldsymbol{D}^{-1}\boldsymbol{a}\boldsymbol{b}^T\right)$$

where

$$\boldsymbol{D} = diag\,(P_1(\lambda), \dots, P_n(\lambda)).$$

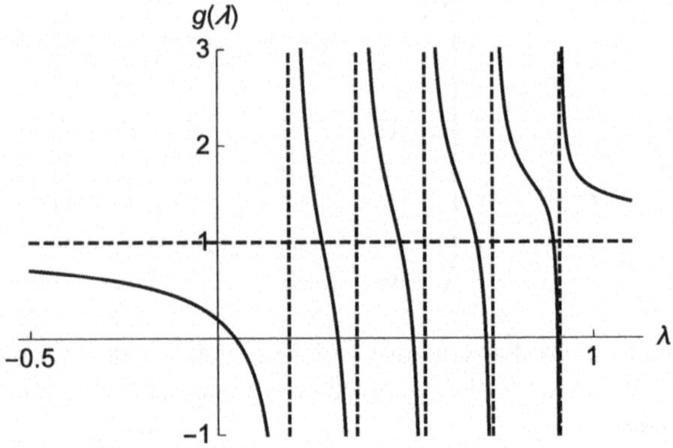

Fig. 4.1 Graph of $g(\lambda)$ with $S = 5$ for $\gamma > 0$

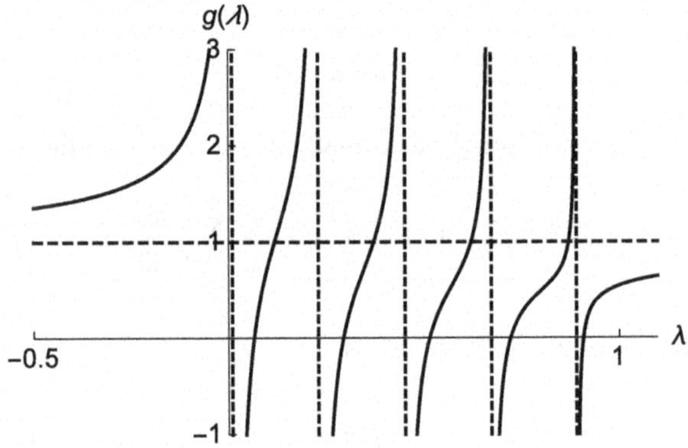

Fig. 4.2 Graph of $g(\lambda)$ with $S = 5$ for $\gamma < 0$

So

$$\varphi(\lambda) = \prod_{k=1}^{n} \left(\lambda^2 - \lambda(1 - K_k) - \frac{K_k \gamma}{2} \right) \left[1 + \sum_{k=1}^{n} \frac{\frac{K_k \gamma}{2}}{\lambda^2 - \lambda(1 - K_k) - \frac{K_k \gamma}{2}} \right]. \tag{4.41}$$

The roots of the quadratic polynomials in the first factor are real and between -1 and $+1$ since it is easy to see that the conditions (C9) of Appendix C are satisfied. We can first show the following simple sufficient stability condition.

Proposition 4.2 *Assume that*

$$|\gamma| < \frac{2}{n-1},$$

then all eigenvalues are inside the unit circle implying the asymptotic stability of the equilibrium with model (4.39).

Proof Let $|v_k| = \max |v_j|$, then

$$\left| \sum_{i \neq k} \frac{K_k \gamma}{2} v_i \right| \leq K_k (n-1) \frac{|\gamma|}{2} |v_k| < K_k |v_k| .$$

Therefore, the above quadratic eigenvalue equation implies that

$$\lambda^2 - (1 - K_k)\,\lambda + K_k \Gamma_k = 0$$

where $|\Gamma_k| < 1$. From conditions (C9) of Appendix C, we know that the roots are inside the unit circle if and only if

$$K_k \Gamma_k < 1$$

$$1 - K_k + K_k \Gamma_k + 1 > 0$$

$$-1 + K_k + K_k \Gamma_k + 1 > 0$$

which are all satisfied, since the last two can be written as

$$K_k (1 - \Gamma_k) < 2 \text{ and } K_k (1 + \Gamma_k) > 0,$$

and the first inequality is also clearly satisfied. ■

The other eigenvalues are the solutions of equation

$$g(\lambda) = 1 + \sum_{k=1}^{n} \frac{\frac{K_k \gamma}{2}}{\lambda^2 - \lambda (1 - K_k) - \frac{K_k \gamma}{2}} = 0. \tag{4.42}$$

We cannot use the same idea about the location of the solutions as before, since there is the possibility of complex roots.

Proposition 4.3 *The equilibrium of the linear oligopoly model* (4.39) *with one period delay is locally asymptotically stable if all solutions of equation* (4.42) *are inside the unit circle.*

If the firms select identical speed of adjustment, then $K_1 = K_2 = \cdots = K_n = K$ and Eq. (4.42) can be rewritten as

$$\lambda^2 - \lambda (1 - K) + \frac{(n-1) K \gamma}{2} = 0.$$

The roots are inside the unit circle if and only if

$$\frac{(n-1) K \gamma}{2} < 1,$$

$$1 - K + \frac{(n-1)K\gamma}{2} + 1 > 0$$

$$-1 + K + \frac{(n-1)K\gamma}{2} + 1 > 0$$

which can be rewritten as

$$-\frac{2}{n-1} < \gamma < \frac{2}{(n-1)\,K}.$$

So, if $\gamma > 0$, then it is $\gamma < 2/[(n-1)\,K]$ and if $\gamma < 0$, then it is $\gamma > -2/(n-1)$. Notice that at $K = 1$, these conditions are the same as those obtained for the system without delays.

We turn our attention to hyperbolic oligopolies. For the sake of simplicity, no product differentiation is assumed, that is, $\Theta_{kj} = 1$ for all k and j. Then $Q_k = \sum_{i \neq k}^{n} q_i$ and from (4.18), the marginal profit of firm k is

$$\frac{\partial \pi_k}{\partial q_k} = \frac{Q_k}{(q_k + Q_k)^2} - c_k \tag{4.43}$$

implying that the best response of firm k is given as

$$R_k\,(Q_k) = \sqrt{\frac{Q_k}{c_k}} - Q_k \tag{4.44}$$

by assuming interior optimum. This value is positive if $c_k Q_k < 1$, otherwise the best response is zero. We can ignore such firms and decrease the number of firms by one. From (4.44), we see that at the equilibrium,

$$q_k + Q_k = Q = \sqrt{\frac{Q_k}{c_k}}$$

where $Q = \sum_{k=1}^{n} q_k$ is the industry output. That is,

$$c_k Q^2 = Q_k,$$

and by adding this relation for all k, we have

$$CQ^2 = (n-1)Q$$

where

$$C = \sum_{k=1}^{n} c_k$$

implying that

$$Q = \frac{n-1}{C} \text{ and } Q_k = c_k \left(\frac{n-1}{C}\right)^2.$$

The equilibrium outputs of the firms are therefore

$$\begin{aligned} q_k &= Q - Q_k, \\ &= \frac{n-1}{C} - c_k \left(\frac{n-1}{C}\right)^2, \\ &= \frac{(n-1)\left[C-(n-1)c_k\right]}{C^2}. \end{aligned} \tag{4.45}$$

Notice that this is positive if

$$c_k < \frac{1}{n-1} \sum_{i=1}^{n} c_i$$

meaning that c_k cannot be much larger than the average marginal cost of the firms.

Assume first that the firms select partial adjustment toward best responses, then the following dynamics is obtained:

$$q_k(t+1) = (1-K_k)\, q_k(t) + K_k \left(\sqrt{\frac{\sum_{i \neq k}^{n} q_i(t)}{c_k}} - \sum_{i \neq k}^{n} q_i(t) \right). \tag{4.46}$$

Notice that at the equilibrium for all $i \neq k$

$$\frac{\partial q_k(t+1)}{\partial q_i(t)} = K_k \left(\frac{1}{2} \sqrt{\frac{1}{c_k Q_k}} - 1 \right) = K_k \left(\frac{C}{2c_k(n-1)} - 1 \right) = K_k \gamma_k$$

where $\gamma_k > -1$ is the derivative of $R_k(Q_k)$. Notice that there is no upper bound for γ_k, since if c_k is much smaller than C, then γ_k can become arbitrarily large.

The Jacobian of system (4.46) has the special form

$$\boldsymbol{J} = \begin{pmatrix} 1-K_1 & K_1\gamma_1 & \ldots & K_1\gamma_1 \\ K_2\gamma_2 & 1-K_2 & \ldots & K_2\gamma_2 \\ \cdot & \cdot & \ldots & \cdot \\ K_n\gamma_n & K_n\gamma_n & \ldots & 1-K_n \end{pmatrix}.$$

All eigenvalues are inside the unit circle of one norm of this matrix is less than one. The row norm is below one, if for all k,

$$K_k(n-1)\,|\gamma_k| < K_k$$

or

$$|\gamma_k| < \frac{1}{n-1}.$$

The column norm is below unity, if for all k,

$$\sum_{i \neq k}^{n} K_i |\gamma_i| < K_k.$$

These are only simple sufficient conditions. We can also determine the characteristic polynomial using the same method that was used before.

By introducing the notation

$$\boldsymbol{D} = diag\left(1 - K_1\left(1 + \gamma_1\right), \ldots, 1 - K_n\left(1 + \gamma_n\right)\right)$$

$$\boldsymbol{a} = \left(K_1\gamma_1, \ldots, K_n\gamma_n\right)^T$$

and

$$\boldsymbol{b}^T = (1, \ldots, 1),$$

the characteristic polynomial can be given as

$$\begin{aligned}
\varphi(\lambda) &= \det\left(\boldsymbol{D} + \boldsymbol{a}\boldsymbol{b}^T - \lambda \boldsymbol{I}\right) \\
&= \det\left(\boldsymbol{D} - \lambda\boldsymbol{I}\right)\det\left(\boldsymbol{I} + \left(\boldsymbol{D} - \lambda\boldsymbol{I}\right)^{-1}\boldsymbol{a}\boldsymbol{b}^T\right) \\
&= \prod_{k=1}^{n}\left(1 - K_k\left(1 + \gamma_k\right) - \lambda\right)\left[1 + \sum_{k=1}^{n}\frac{K_k\gamma_k}{1 - K_k\left(1 + \gamma_k\right) - \lambda}\right].
\end{aligned} \tag{4.47}$$

From the first factor $\lambda = 1 - K_k\left(1 + \gamma_k\right)$, which is between -1 and $+1$ if and only if

$$-1 < 1 - K_k \frac{C}{2c_k(n-1)} < +1$$

which occurs if

$$c_k > \frac{K_k C}{4(n-1)}. \tag{4.48}$$

The other eigenvalues are solutions of equation

$$g(\lambda) = 1 + \sum_{k=1}^{n} \frac{K_k\gamma_k}{1 - K_k\left(1 + \gamma_k\right) - \lambda} = 0. \tag{4.49}$$

Under condition (4.48), all poles are between -1 and $+1$. However, the signs of γ_k are indeterminate, therefore we cannot derive simple stability conditions as we did in the linear case. However, we can have the following result.

Proposition 4.4 *The equilibrium of the hyperbolic oligopoly with model* (4.46) *is locally asymptotically stable if* (4.48) *holds and the roots of equation* (4.49) *are inside the unit circle.*

Notice that if $c_1 = \cdots = c_n = c$, then for $n \geq 3$,

$$\gamma_1 = \cdots = \gamma_n = -\frac{n-2}{2(n-1)} < 0$$

and if the c_k values are sufficiently close to each other, then all γ_k values remain negative and (4.48) also holds for all k. Then all poles of (4.49) are between -1 and $+1$, $g'(\lambda) < 0$ and the graph of $g(\lambda)$ is similar to that shown in Fig. 4.1. The roots between the poles are clearly between -1 and $+1$, and the remaining root before the smallest pole is larger than -1 if $g(-1) > 0$, which can be written as

$$\sum_{k=1}^{n} \frac{K_k \gamma_k}{2 - K_k (1 + \gamma_k)} > -1.$$

In the symmetric case if in addition, $K_1 = \cdots = K_n = K$ and $c_1 = \cdots = c_n = c$, this condition can be rewritten as

$$nK\gamma > -2 + K(1 + \gamma)$$

since the γ_k values are also the same. That is,

$$\gamma = -\frac{n-2}{2(n-1)} > \frac{K-2}{(n-1)K}$$

or

$$nK < 4.$$

So for $n = 2, 3$ stability is guaranteed, for $n = 4$ only if $K < 1$, and for $n > 4$, when $K < 4/n$.

Assume next that there is a unit delay for the firms in obtaining the output values of the rest of the industry. In this case, we can proceed similarly to the linear case. The dynamics is described again by Eq. (4.39) which is nonlinear since R_k is nonlinear. By introducing the new variables

$$z_k(t) = q_k(t-1),$$

the nonlinear $2n$-dimensional system is obtained

$$
\begin{aligned}
q_k(t+1) &= (1-K_k)q_k(t) + K_k R_k\left(\textstyle\sum_{i\neq k}^{n} z_i(t)\right) \\
z_k(t+1) &= q_k(t)
\end{aligned} \tag{4.50}
$$

with Jacobian matrix

$$
J = \begin{pmatrix}
1-K_1 & 0 & \ldots & 0 & 0 & K_1\gamma_1 & \ldots & K_1\gamma_1 \\
0 & 1-K_2 & \ldots & 0 & K_2\gamma_2 & 0 & \ldots & K_2\gamma_2 \\
\cdot & \cdot & \ldots & \cdot & \cdot & \cdot & \ldots & \cdot \\
0 & 0 & \ldots & 1-K_n & K_n\gamma_n & K_n\gamma_n & \ldots & 0 \\
1 & 0 & \ldots & 0 & 0 & 0 & \ldots & 0 \\
0 & 1 & \ldots & 0 & 0 & 0 & \ldots & 0 \\
\cdot & \cdot & \ldots & \cdot & \cdot & \cdot & \ldots & \cdot \\
0 & 0 & \ldots & 1 & 0 & 0 & \ldots & 0
\end{pmatrix}
$$

where the derivative of $R_k(Q_k)$ at the equilibrium is given as

$$
\gamma_k = \frac{C}{2c_k(n-1)} - 1.
$$

The eigenvalue equations are now

$$
\begin{aligned}
(1-K_k)u_k + \textstyle\sum_{i\neq k}^{n} K_k\gamma_k v_i &= \lambda u_k \\
u_k &= \lambda v_k,
\end{aligned}
$$

so for $k = 1, 2, \ldots, n$,

$$
\left(\lambda^2 - \lambda(1-K_k)\right) v_k - \sum_{i\neq k}^{n} K_k\gamma_k v_i = 0. \tag{4.51}
$$

Similarly to Proposition 4.2, we can prove the following result.

Proposition 4.5 *The equilibrium of model* (4.50) *is locally asymptotically stable if*

$$
|\gamma_k| < \frac{1}{n-1} \text{ for all } k.
$$

Proof Let $|v_k| = \max_{1\leq i\leq n} |v_i|$, then

$$
\left|\sum_{i\neq k}^{n} K_k\gamma_k v_i\right| < (n-1)K_k\frac{1}{n-1}|v_k| = K_k|v_k|,
$$

so (4.51) can be rewritten as

$$\lambda^2 - \lambda(1 - K_k) - \Gamma_k K_k = 0,$$

where $|\Gamma_k| < 1$. All conditions (C9) of Appendix C hold:

$$-\Gamma_k K_k < 1,$$

$$1 - K_k - \Gamma_k K_k + 1 > 0,$$

$$K_k - 1 - \Gamma_k K_k + 1 > 0.$$

The first inequality holds obviously, the second and third can be rewritten as

$$K_k (1 + \Gamma_k) < 2,$$

$$K_k (1 - \Gamma_k) > 0.$$

■

The characteristic polynomial is the determinant of system (4.51), which can be given in matrix form as follows:

$$\det\left(\boldsymbol{D} + \boldsymbol{a}\boldsymbol{b}^T\right) = \det\left(\boldsymbol{D}\right)\det\left(\boldsymbol{I} + \boldsymbol{D}^{-1}\boldsymbol{a}\boldsymbol{b}^T\right)$$

where

$$\boldsymbol{D} = diag\left(\lambda^2 - \lambda(1 - K_1) + K_1\gamma_1, \ldots, \lambda^2 - \lambda(1 - K_n) + K_n\gamma_n\right)$$

$$\boldsymbol{a} = (-K_1\gamma_1, \ldots, -K_n\gamma_n)^T$$

and

$$\boldsymbol{b}^T = (1, \ldots, 1)\,.$$

So

$$\varphi(\lambda) = \prod_{k=1}^{n}\left(\lambda^2 - \lambda(1 - K_k) + K_k\gamma_k\right)\left[1 + \sum_{k=1}^{n}\frac{-K_k\gamma_k}{\lambda^2 - \lambda(1 - K_k) + K_k\gamma_k}\right]. \tag{4.52}$$

We have the same difficulty as in the case of linear oligopolies, since there is the possibility again for the occurrence of complex roots.

Proposition 4.6 *The equilibrium of the hyperbolic oligopoly model* (4.50) *with one period delay is locally asymptotically stable if all roots of equations*

$$\lambda^2 - \lambda(1 - K_k) + K_k\gamma_k = 0 \tag{4.53}$$

$$1+\sum_{k=1}^{n}\frac{-K_k\gamma_k}{\lambda^2-\lambda(1-K_k)+K_k\gamma_k}=0 \tag{4.54}$$

are inside the unit cycle. The roots of (4.53) are inside the unit circle if and only if

$$-1<\gamma_k<\frac{1}{K_k}$$

where the left hand side is always satisfied.

If $K_1=\cdots=K_n=K$, then Eq. (4.54) has the form

$$\lambda^2-\lambda(1-K)-(n-1)K\gamma_k=0,$$

the roots of which are inside the unit circle if and only if

$$-(n-1)K\gamma_k<1,$$

$$-1+K-(n-1)K\gamma_k+1>0,$$

$$1-K-(n-1)K\gamma_k+1>0,$$

which is the case if and only if

$$-\frac{1}{(n-1)K}<\gamma_k<\frac{1}{n-1}.$$

Therefore the combined stability condition is as follows:

$$\max\left\{-\frac{1}{(n-1)K},-1\right\}<\gamma_k<\frac{1}{n-1},$$

and in the special case of $K=1$, the condition is

$$-\frac{1}{n-1}<\gamma_k<\frac{1}{n-1}.$$

This is the same as the sufficient condition without delay obtained from examining the row norm of the Jacobian. Notice that the lower and upper bounds are the half values of those in the linear case.

We now turn our attention to Bertrand oligopolies. Consider first the linear case. From (4.7), we already derived that the marginal profit of firm k is

$$\frac{\partial\bar{\pi}_k(p_1,\ldots,p_n)}{\partial p_k}=\frac{-[1+(n-2)\gamma](p_k-c_k)+[1+(n-2)\gamma](\alpha_k-p_k)-\gamma\sum_{i\neq k}^{n}(\alpha_i-p_i)}{(1-\gamma)[1+(n-1)\gamma]}$$

Fig. 4.3 Graph of $\bar{\gamma}$ as function of γ

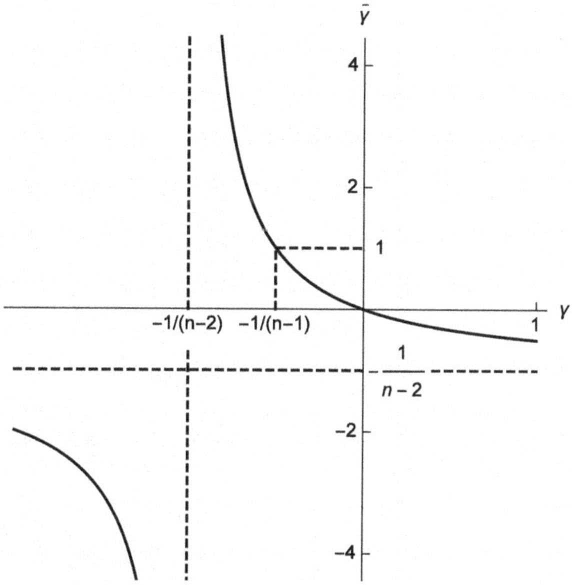

and equating the numerator with zero, we have the best response of firm k with the notation $\boldsymbol{p}_{-k} = (p_1, .., p_{k-1}, p_{k+1}, \ldots, p_n)$,

$$R_k\left(\boldsymbol{p}_{-k}\right) = \frac{\alpha_k + c_k}{2} - \frac{\gamma}{2\left(1 + (n-2)\gamma\right)} \sum_{i \neq k}^{n} \left(\alpha_i - p_i\right), \tag{4.55}$$

therefore the dynamics with assuming partial adjustments toward best responses is described with equations

$$p_k(t+1) = \left(1 - K_k\right) p_k(t) + K_k R_k\left(\boldsymbol{p}_{-k}(t)\right). \tag{4.56}$$

The Jacobian of this system has the same form as that of system (4.32) with the only difference that γ is now replaced by

$$\bar{\gamma} = -\frac{\gamma}{1 + (n-2)\gamma}.$$

Notice that $\bar{\gamma}$ locally strictly decreases in γ; however, at $\gamma = -1/(n-2)$, it has a pole with $+\infty$ and $-\infty$ limits at the two sides of the pole. The graph of $\bar{\gamma}$ as function of γ is shown in Fig. 4.3.

Consider now the characteristic equation (4.35). The eigenvalues

$$\lambda = 1 - K_j\left(1 - \frac{\bar{\gamma}}{2}\right)$$

are between -1 and $+1$, if

$$2 - \frac{4}{K_j} < \bar{\gamma} < 2.$$

Since we have eliminated the region

$$-\frac{1}{n-2} < \gamma < -\frac{1}{n-1}$$

from consideration, we have to assume that $\bar{\gamma} \leq 1$, so this condition becomes

$$2 - \frac{4}{K_j} < \bar{\gamma} < 1.$$

In terms of γ, this can be rewritten for $n \geq 3$ as either

$$-\frac{1}{n-1} < \gamma \leq 1 \text{ or } -1 \leq \gamma \leq \frac{2K_j - 4}{(4n-8) - K_j(2n-3)}$$

where the right hand side is larger than -1 if $n > 3$. Then Proposition 4.1 remains valid with γ being replaced by $\bar{\gamma}$.

In the special case of $K_1 = \cdots = K_n = 1$, the stability condition was shown to be

$$-\frac{2}{n-1} < \bar{\gamma} < \frac{2}{n-1}$$

and it is easy to prove that this is the case when $n > 3$ and either

$$-1 \leq \gamma < -\frac{2}{n-3} \text{ or } -\frac{2}{3n-5} < \gamma \leq 1$$

where the first interval is nonempty for $n > 5$ and the left hand side of the second inequality is above $-1/(n-1)$ for $n > 3$.

Assume next that there is a unit time delay for the firms to obtain data about the output of the rest of the industry. Then the model is identical with (4.39), which can be rewritten as a $2n$-dimensional no-delay system with characteristic polynomial (4.41), where γ is replaced again with $\bar{\gamma}$. Equation

$$\lambda^2 - \lambda(1 - K_k) - \frac{K_k \bar{\gamma}}{2} = 0$$

has its roots inside the unit circle if and only if

$$-\frac{2}{K_k} < \bar{\gamma}$$

which occurs if for $n > 3$, either

$$-1 < \gamma < -\frac{2}{2n-4-K_k} \text{ or } -\frac{1}{n-1} < \gamma \leq 1.$$

For $n > 3$, the first interval is nonempty. We also consider the concavity of the payoff function. Then Proposition 4.3 remains valid if γ is replaced with $\bar{\gamma}$ with these addition conditions.

The case of hyperbolic oligopolies can be discussed similarly. Based on payoff function (4.21), the best response of firm k can be obtained by equating the marginal profit to zero to get equation

$$p_k = R_k\left(\boldsymbol{p}_{-k}\right) = \sqrt{-\frac{c_k b_{kk}^{-1}}{\sum_{j\neq k}^{n} b_{kj}^{-1}\dfrac{1}{p_j}}}. \tag{4.57}$$

Dynamics with assumption of partial adjustments toward best responses results in the dynamic equation

$$p_k(t+1) = (1-K_k)\,p_k(t) + K_k R_k\left(\boldsymbol{p}_{-k}(t)\right) \tag{4.58}$$

with Jacobian matrix

$$\boldsymbol{J} = \begin{pmatrix} 1-K_1 & K_1\gamma_{12} & \cdots & K_1\gamma_{1n} \\ K_2\gamma_{21} & 1-K_2 & \cdots & K_2\gamma_{2n} \\ \cdot & \cdot & \cdot & \cdot \\ \cdot & \cdot & \cdot & \cdot \\ \cdot & \cdot & \cdot & \cdot \\ K_n\gamma_{n1} & K_n\gamma_{n2} & \cdots & 1-K_n \end{pmatrix}$$

where γ_{kj} is the partial derivative of (4.57) with respect to p_j. In this general form, no result can be derived about the location of the eigenvalues. However, we can have the following simple results.

Assume first that for all k,

$$\sum_{j\neq k}^{n} \left|\gamma_{kj}\right| < 1.$$

Then the row norm of $\boldsymbol{J}$ is less than one, so the eigenvalues are inside the unit circle, implying the local asymptotical stability of the equilibrium with (4.58). The same result holds, if for all k,

$$\sum_{j\neq k}^{n} K_j\left|\gamma_{jk}\right| < K_k,$$

where the column norm of $\boldsymbol{J}$ becomes less than 1.

In the general case, we cannot derive a closed form representation of the characteristic equation. The equilibrium with model (4.58) is locally asymptotically stable if all eigenvalues of matrix $\boldsymbol{J}$ are inside the unit circle.

Consider next the symmetric case where $K_1 = \cdots = K_n = K,\ c_1 = \cdots = c_n = c$ and $\theta_{kj} = \theta$ for all k and j. Then in Eq. (4.19), the coefficient matrix is

$$\boldsymbol{B} = \begin{pmatrix} 1\ \theta\ \ldots\ \theta \\ \theta\ 1\ \ldots\ \theta \\ \cdot\ \ \cdot\ \ \ldots\ \cdot \\ \theta\ \theta\ \ldots\ 1 \end{pmatrix}$$

with inverse $\boldsymbol{B}^{-1} = \left(b_{ij}^{-1}\right)$, where

$$b_{kk}^{-1} = \frac{1+(n-2)\theta}{1+(n-2)\theta-(n-1)\theta^2} \text{ and } b_{kj}^{-1} = -\frac{\theta}{1+(n-2)\theta-(n-1)\theta^2}.$$

Notice that the denominator is $(1-\theta)\left[(n-1)\theta+1\right]$ which is positive if $\theta < 1$. The best response function of firm k is given by Eq. (4.57), which is specialized as follows:

$$p_k = R\left(\boldsymbol{p}_{-k}\right) = \sqrt{\frac{c\left(1+(n-2)\theta\right)}{\theta\sum_{j\neq k}^{n}\dfrac{1}{p_j}}}.$$

At a symmetric equilibrium $p_1 = \cdots = p_n = p^*$,

$$p^* = \sqrt{\frac{c\left(1+(n-2)\theta\right)p^*}{\theta(n-1)}}$$

implying that

$$p^* = \frac{c\left(1+(n-2)\theta\right)}{\theta(n-1)}.$$

In addition,

$$\frac{\partial R\left(\boldsymbol{p}_{-k}\right)}{\partial p_j} = \frac{1}{2p_j^2}\sqrt{\frac{c\left(1+(n-2)\theta\right)}{\theta\left(\sum_{j\neq k}^{n}\dfrac{1}{p_j}\right)^3}}$$

which becomes

$$\bar{\gamma} = \frac{1}{2(n-1)}$$

at the equilibrium. The Jacobian becomes the same as that of system (4.32) where $-\gamma/2$ is replaced by $\bar{\gamma} > 0$. So from (4.38), the stability condition is

$$-2\bar{\gamma} > -\frac{2}{n-1}$$

or

$$\bar{\gamma} < \frac{1}{n-1},$$

which holds. Therefore, the symmetric equilibrium is locally asymptotically stable.

Assume next that there is a unit delay in obtaining information about the prices of the competitors. Then the dynamic equation (4.58) is modified as

$$p_k(t+1) = (1-K_k)\, p_k(t) + K_k R_k \left(\boldsymbol{p}_{-k}(t-1)\right) \tag{4.59}$$

and by introducing the new variables

$$z_j(t) = p_j(t-1),$$

we have the $2n$-dimensional system

$$p_k(t+1) = (1-K_k)\, p_k(t) + K_k R_k \left(z_{-k}(t)\right),$$

$$z_k(t+1) = p_k(t)$$

with Jacobian

$$\boldsymbol{J} = \begin{pmatrix} 1-K_1 & 0 & \dots & 0 & 0 & K_1\gamma_{12} & \dots & K_1\gamma_{1n} \\ 0 & 1-K_2 & \dots & 0 & K_2\gamma_{21} & 0 & \dots & K_2\gamma_{2n} \\ \cdot & \cdot & \dots & \cdot & \cdot & \cdot & \dots & \cdot \\ 0 & 0 & \dots & 1-K_n & K_n\gamma_{n1} & K_n\gamma_{n2} & \dots & 0 \\ 1 & 0 & \dots & 0 & 0 & 0 & \dots & 0 \\ 0 & 1 & \dots & 0 & 0 & 0 & \dots & 0 \\ \cdot & \cdot & \dots & \cdot & \cdot & \cdot & \dots & \cdot \\ 0 & 0 & \dots & 1 & 0 & 0 & \dots & 0 \end{pmatrix}.$$

The eigenvalue equations have the form similar to the previous case of model (4.40):

$$(1-K_k)\, u_k + K_k \sum_{j\neq k}^{n} \gamma_{kj} v_j = \lambda u_k$$

$$u_k = \lambda v_k$$

and by combining these equations,

$$\left(\lambda^2 - (1 - K_k)\,\lambda\right) v_k - K_k \sum_{j \neq k}^{n} \gamma_{kj} v_j = 0 \tag{4.60}$$

so the characteristic equation can be written as

$$\varphi(\lambda) = \det \begin{pmatrix} P_1(\lambda) & -K_1\gamma_{12} & \dots & -K_1\gamma_{1n} \\ -K_2\gamma_{21} & P_2(\lambda) & \dots & -K_2\gamma_{2n} \\ \cdot & \cdot & \dots & \cdot \\ -K_n\gamma_{n1} & -K_n\gamma_{n2} & \dots & P_n(\lambda) \end{pmatrix} = 0$$

where $P_k(\lambda) = \lambda^2 - (1 - K_k)\,\lambda$. Similarly to the previous case, no result can be derived in general. However, the earlier discussed sufficient conditions can be extended as follows.

Proposition 4.7 *Assume that for all k,*

$$\sum_{j \neq k}^{n} \left|\gamma_{kj}\right| < 1.$$

Then all eigenvalues from (4.60) *are inside the unit circle, implying the local asymptotical stability of the equilibrium with* (4.59).

Proof Let k be selected as $|v_k| == \max_{1 \le j \le n} \left|v_j\right|$, then

$$\left|\sum_{j \neq k}^{n} \gamma_{kj} v_j\right| \le \sum_{j \neq k}^{n} \left|\gamma_{kj}\right| \left|v_j\right| \le \sum_{j \neq k}^{n} \left|\gamma_{kj}\right| \left|v_k\right|.$$

Therefore in Eq. (4.60), we may write

$$\sum_{j \neq k}^{n} \gamma_{kj} v_j = \Gamma_k v_k$$

with $|\Gamma_k| < 1$. Equation (4.60) can be now rewritten as

$$\lambda^2 - (1 - K_k)\,\lambda - \Gamma_k K_k = 0. \tag{4.61}$$

The roots of this equation are inside the unit circle. Conditions (C9) of Appendix C are as follows:

$$-\Gamma_k K_k < 1,$$

$$-1 + K_k - \Gamma_k K_k + 1 > 0,$$

$$1 - K_k - \Gamma_k K_k + 1 > 0.$$

They can be simplified as

$$\Gamma_k K_k > -1,$$

$$K_k (1 - \Gamma_k) > 0,$$

$$K_k (1 + \Gamma_k) < 2.$$

These relations hold as $0 < K_k \leq 1$ and $-1 < \Gamma_k < 1$. ■

As an example of mixed oligopolies, we will now discuss the linear case. As in the Introduction of this chapter, assume that firms k $(1 \leq k \leq r)$ are quantity setting and firms k $(k = r + 1, \ldots, n)$ are price setting, where $1 \leq r \leq n - 1$. Using the notation of Eqs. (4.8) and (4.9), $\boldsymbol{B}^{KK}$ is $r \times r$ with unit diagonal elements and γ off-diagonal elements. Similarly, $\boldsymbol{B}^{\bar{K}\bar{K}}$ is $(n - r) \times (n - r)$ with the same diagonal and off-diagonal elements. In addition, all elements of the $r \times (n - r)$ type $\boldsymbol{B}^{K\bar{K}}$ matrix and those of the $(n - r) \times r$ type $\boldsymbol{B}^{\bar{K}K}$ matrix are equal to γ. The inverses of $\boldsymbol{B}^{KK}$ and $\boldsymbol{B}^{\bar{K}\bar{K}}$ have the same structure as $\boldsymbol{B}^{-1}$ in Eq. (4.6) with the only difference that n is replaced by r and $n - r$, respectively. By usual simple matrix algebra based on Eqs. (4.11) and (4.12) we have that for $1 \leq k \leq r$,

$$p_k = \frac{[1+(n-r-1)\gamma]\alpha_k - \gamma \sum_{i=r+1}^{n} \alpha_i - (1-\gamma)[1+(n-r)\gamma]q_k + \gamma\left(\sum_{i=r+1}^{n} p_i - (1-\gamma)\sum_{i=1, i\neq k}^{r} q_i\right)}{1+(n-r-1)\gamma} \tag{4.62}$$

and for $r + 1 \leq k \leq n$,

$$q_k = \frac{[1+(n-r-1)\gamma]\alpha_k - \gamma \sum_{i=r+1}^{n} \alpha_i - [1+(n-r-2)\gamma]p_k + \gamma\left(\sum_{i=r+1, i\neq k}^{n} p_i - (1-\gamma)\sum_{i=1}^{r} q_i\right)}{(1-\gamma)[1+(n-r-1)\gamma]}. \tag{4.63}$$

The profit functions of the firms are given as follows:

$$\bar{\pi}'_k \left(\boldsymbol{q}^K, \boldsymbol{p}^{\bar{K}}\right) = (p_k - c_k)\, q_k$$

where for $k = 1, \ldots, r$, p_k is given from (4.62) and for $k = r + 1, \ldots, n$, q_k is given from (4.63). The marginal profits are obtained as

$$\frac{\partial \bar{\pi}'_k}{\partial q_k} = \frac{\partial p_k}{\partial q_k} q_k + (p_k - c_k) \text{ for } 1 \leq k \leq r$$

and

$$\frac{\partial \bar{\pi}'_k}{\partial p_k} = q_k + (p_k - c_k) \frac{\partial q_k}{\partial p_k} \text{ for } r + 1 \leq k \leq n.$$

Substituting (4.62) for p_k in the first equation, (4.63) for q_k in the second equation and equating the marginal profits with zero, the best responses are obtained:

for $k = 1, 2, \ldots, r$,

$$q_k = R_k\left(\boldsymbol{q}_{-k}^{K}, \boldsymbol{p}^{\bar{K}}\right)$$

$$= \frac{[1+(n-r-1)\gamma]\alpha_k - \gamma\sum_{i=r+1}^{n}\alpha_i + \gamma\left(\sum_{i=r+1}^{n} p_i - (1-\gamma)\sum_{i=1,\ i\neq k}^{r} q_i\right)}{2(1-\gamma)(1+(n-r)\gamma)}$$

and for $k = r+1, \ldots, n$,

$$p_k = R_k\left(\boldsymbol{q}^{K}, \boldsymbol{p}_{-k}^{\bar{K}}\right)$$

$$= \frac{[1+(n-r-1)\gamma]\alpha_k - \gamma\sum_{i=r+1}^{n}\alpha_i + \gamma\left(\sum_{i=r+1, i\neq k}^{n} p_i - (1-\gamma)\sum_{i=1}^{r} q_i\right)}{2(1+(n-r-2)\gamma)}.$$

The dynamic system with partial adjustment toward best responses has the form:

$$\begin{aligned} q_k(t+1) &= (1-K_k)q_k(t) + K_k R_k\left(\boldsymbol{q}_{-k}^{K}(t), \boldsymbol{p}^{\bar{K}}(t)\right) \text{ for } 1 \leq k \leq r, \\ p_k(t+1) &= (1-K_k)p_k(t) + K_k R_k\left(\boldsymbol{q}^{K}(t), \boldsymbol{p}_{-k}^{\bar{K}}(t)\right) \text{ for } r+1 \leq k \leq n. \end{aligned} \tag{4.64}$$

with Jacobian matrix

$$\boldsymbol{J} = \begin{pmatrix} \boldsymbol{J}_{11} & \boldsymbol{J}_{12} \\ \boldsymbol{J}_{21} & \boldsymbol{J}_{22} \end{pmatrix}$$

where the diagonal elements of $\boldsymbol{J}_{11}$ are $1 - K_k$ $(1 \leq k \leq r)$ and the off-diagonal elements are $K_k\gamma_1$ with

$$\gamma_1 = -\frac{\gamma}{2\left(1+(n-r)\gamma\right)}.$$

All elements of $\boldsymbol{J}_{12}$ are equal to $K_k\gamma_2$ with

$$\gamma_2 = \frac{\gamma}{2(1-\gamma)\left(1+(n-r)\gamma\right)},$$

all elements of $\boldsymbol{J}_{21}$ are equal to $K_k\gamma_3$ with

$$\gamma_3 = -\frac{\gamma(1-\gamma)}{2\left(1+(n-r-2)\gamma\right)}$$

and the diagonal elements of $\boldsymbol{J}_{22}$ are $1 - K_k$ $(r+1 \leq k \leq n)$ and the off-diagonal elements are $K_k\gamma_4$ where

$$\gamma_4 = \frac{\gamma}{2\left(1+(n-r-2)\gamma\right)}.$$

So the eigenvalue equations have the form

$$(1-K_k)\,u_k + K_k\gamma_1 \textstyle\sum_{i=1,i\neq k}^{r} u_i + K_k\gamma_2 \sum_{j=r+1}^{n} u_j = \lambda u_k \quad (1 \le k \le r)\,,$$

$$K_k\gamma_3 \textstyle\sum_{i=1}^{r} u_i + (1-K_k)\,u_k + K_k\gamma_4 \sum_{j=r+1,j\neq k}^{n} u_j = \lambda u_k \quad (r+1 \le k \le n)\,.$$

Let

$$U = \sum_{k=1}^{r} u_k \text{ and } \bar{U} = \sum_{k=r+1}^{n} u_k,$$

then we have

$$(1-K_k-K_k\gamma_1-\lambda)\,u_k + K_k\gamma_1 U + K_k\gamma_2\bar{U} = 0 \quad (1 \le k \le r)\,,$$

$$K_k\gamma_3 U + (1-K_k-K_k\gamma_4-\lambda)\,u_k + K_k\gamma_4\bar{U} = 0 \quad (r+1 \le k \le n)\,.$$

So for $1 \le k \le r$,

$$u_k = -\frac{K_k}{1-K_k\,(1+\gamma_1)-\lambda}\left(\gamma_1 U + \gamma_2\bar{U}\right) \tag{4.65}$$

and for $r+1 \le k \le n$,

$$u_k = -\frac{K_k}{1-K_k\,(1+\gamma_4)-\lambda}\left(\gamma_3 U + \gamma_4\bar{U}\right). \tag{4.66}$$

Adding Eq. (4.65) for $k = 1, 2, \ldots, r$ and Eq. (4.66) for $k = r+1, \ldots, n$, we have

$$U = \left(\gamma_1 U + \gamma_2\bar{U}\right) \textstyle\sum_{k=1}^{r} \left(-\dfrac{K_k}{1-K_k\,(1+\gamma_1)-\lambda}\right),$$

$$\bar{U} = \left(\gamma_3 U + \gamma_4\bar{U}\right) \textstyle\sum_{k=r+1}^{n} \left(-\dfrac{K_k}{1-K_k\,(1+\gamma_4)-\lambda}\right).$$

We have two homogeneous linear equations for unknowns U and $\bar{U}$ with coefficient matrix

$$\begin{pmatrix} 1+\gamma_1 \sum_{k=1}^{r} \dfrac{K_k}{1-K_k\,(1+\gamma_1)-\lambda} & \gamma_2 \sum_{k=1}^{r} \dfrac{K_k}{1-K_k\,(1+\gamma_1)-\lambda} \\ \gamma_3 \sum_{k=r+1}^{n} \dfrac{K_k}{1-K_k\,(1+\gamma_4)-\lambda} & 1+\gamma_4 \sum_{k=r+1}^{n} \dfrac{K_k}{1-K_k\,(1+\gamma_4)-\lambda} \end{pmatrix}.$$

There are two possibilities. If $U = \bar{U} = 0$, then for all k $(1 \le k \le r)$, either $u_k = 0$ or $\lambda = 1 - K_k(1+\gamma_1)$. In the second case, λ is between -1 and $+1$, if $-1 < \gamma_1 < \frac{2}{K_k} - 1$. For $k = r+1, \ldots, n$, the value of u_k is zero or $\lambda = 1 - K_k(1+\gamma_4)$, in which case λ is between -1 and $+1$, if $-1 < \gamma_4 < \frac{2}{K_k} - 1$. If λ is not equal to any of these values, then for all k, $u_k = 0$, which is impossible, since eigenvectors must be nonzero. If at least one of U and $\bar{U}$ is nonzero, then the determinant of the coefficient matrix is zero, that is

$$1 + \gamma_1 \sum_{k=1}^{r} \frac{K_k}{1 - K_k(1+\gamma_1) - \lambda} + \gamma_4 \sum_{k=r+1}^{n} \frac{K_k}{1 - K_k(1+\gamma_4) - \lambda}$$

$$+ (\gamma_1\gamma_4 - \gamma_2\gamma_3)\left(\sum_{k=1}^{r} \frac{K_k}{1 - K_k(1+\gamma_1) - \lambda}\right)\left(\sum_{k=r+1}^{n} \frac{K_k}{1 - K_k(1+\gamma_4) - \lambda}\right) = 0.$$

Since

$$\gamma_1\gamma_4 - \gamma_2\gamma_3 = \frac{-\gamma^2}{4[1+(n-r)\gamma][1+(n-r-2)\gamma]} + \frac{\gamma^2(1-\gamma)}{4(1-\gamma)[1+(n-r)\gamma][1+(n-r-2)\gamma]} = 0,$$

the characteristic equation has the form

$$1 + \gamma_1 \sum_{k=1}^{r} \frac{K_k}{1 - K_k(1+\gamma_1) - \lambda} + \gamma_4 \sum_{k=r+1}^{n} \frac{K_k}{1 - K_k(1+\gamma_4) - \lambda} = 0 \tag{4.67}$$

which is equivalent to a polynomial equation of degree n.

Proposition 4.8 *The equilibrium with model* (4.64) *is asymptotically stable if all roots of equation* (4.67) *are inside the unit circle and both γ_1 and γ_4 are between*

$$-1 \textit{ and } \frac{2}{K_k} - 1 \textit{ for all } k.$$

Because of the possibility of complex roots, it is impossible to give simple conditions.

Assume next that in determining best responses the firms use delayed quantity and price information, where the length of the delay is one time period. Then system (4.64) is modified as

$$q_k(t+1) = (1-K_k)q_k(t) + K_k R_k\left(\boldsymbol{q}_{-k}^{K}(t-1), \boldsymbol{p}^{\bar{K}}(t-1)\right) \text{ for } 1 \le k \le r,$$

$$p_k(t+1) = (1-K_k)p_k(t) + K_k R_k\left(\boldsymbol{q}^{K}(t-1), \boldsymbol{p}_{-k}^{\bar{K}}(t-1)\right) \text{ for } r+1 \le k \le n. \tag{4.68}$$

Introducing the new variables

$$z_k^K(t) = q_k(t-1) \text{ for } 1 \le k \le r \text{ and } z_k^{\bar{K}}(t) = p_k(t-1) \text{ for } r+1 \le k \le n,$$

a $2n$-dimensional system is obtained,

$$\begin{aligned}
q_k(t+1) &= (1-K_k)q_k(t) + K_k R_k\left(z^K{}_k(t), z^{\bar{K}}(t)\right) && (1 \le k \le r)\\
p_k(t+1) &= (1-K_k)p_k(t) + K_k R_k\left(z^K(t), z^{\bar{K}}_{-k}(t)\right) && (r+1 \le k \le n)\\
z_k^K(t+1) &= q_k(t) && (1 \le k \le r)\\
z_k^{\bar{K}}(t+1) &= p_k(t) && (r+1 \le k \le n)
\end{aligned}$$

with Jacobian matrix

$$\boldsymbol{J} = \begin{pmatrix} \boldsymbol{J}_{11} & \boldsymbol{J}_{12} & \boldsymbol{J}_{13} & \boldsymbol{J}_{14} \\ \boldsymbol{J}_{21} & \boldsymbol{J}_{22} & \boldsymbol{J}_{23} & \boldsymbol{J}_{24} \\ \boldsymbol{J}_{31} & \boldsymbol{J}_{32} & \boldsymbol{J}_{33} & \boldsymbol{J}_{34} \\ \boldsymbol{J}_{41} & \boldsymbol{J}_{42} & \boldsymbol{J}_{43} & \boldsymbol{J}_{44} \end{pmatrix}$$

where the sub-matrices in the first row of $\boldsymbol{J}$ are

$$\boldsymbol{J}_{11} = \begin{pmatrix} 1-K_1 & 0 & \cdot & 0 \\ 0 & 1-K_2 & \cdot & 0 \\ \cdot & \cdot & \cdot & \cdot \\ 0 & 0 & \cdot & 1-K_r \end{pmatrix}_{r\times r},$$

$$\boldsymbol{J}_{12} = \boldsymbol{0}_{r\times(n-r)},$$

$$\boldsymbol{J}_{13} = \begin{pmatrix} 0 & K_1\gamma_1 & \cdot & K_1\gamma_1 \\ K_2\gamma_1 & 0 & \cdot & K_2\gamma_1 \\ \cdot & \cdot & \cdot & \cdot \\ K_r\gamma_1 & K_r\gamma_1 & \cdot & 0 \end{pmatrix}_{r\times r},$$

$$\boldsymbol{J}_{14} = \begin{pmatrix} K_1\gamma_2 & K_1\gamma_2 & \cdot & K_1\gamma_2 \\ K_2\gamma_2 & K_2\gamma_2 & \cdot & K_2\gamma_2 \\ \cdot & \cdot & \cdot & \cdot \\ K_r\gamma_2 & K_r\gamma_2 & \cdot & K_r\gamma_2 \end{pmatrix}_{r\times(n-r)},$$

the sub-matrices in the second row are

$$J_{21} = \mathbf{0}_{(n-r)\times r},$$

$$J_{22} = \begin{pmatrix} 1-K_{r+1} & 0 & \cdot & 0 \\ 0 & 1-K_{r+2} & \cdot & 0 \\ \cdot & \cdot & \cdot & \cdot \\ 0 & 0 & \cdot & 1-K_n \end{pmatrix}_{(n-r)\times(n-r)}$$

$$J_{23} = \begin{pmatrix} K_{r+1}\gamma_3 & K_{r+1}\gamma_3 & \cdot & K_{r+1}\gamma_3 \\ K_{r+2}\gamma_3 & K_{r+2}\gamma_3 & \cdot & K_{r+2}\gamma_3 \\ \cdot & \cdot & \cdot & \cdot \\ K_n\gamma_3 & K_n\gamma_3 & \cdot & K_n\gamma_3 \end{pmatrix}_{(n-r)\times r},$$

$$J_{24} = \begin{pmatrix} 0 & K_{r+1}\gamma_4 & \cdot & K_{r+1}\gamma_4 \\ K_{r+2}\gamma_4 & 0 & \cdot & K_{r+2}\gamma_4 \\ \cdot & \cdot & \cdot & \cdot \\ K_n\gamma_4 & K_n\gamma_4 & \cdot & 0 \end{pmatrix}_{(n-r)\times(n-r)},$$

the sub-matrices in the third row are

$$J_{31} = I_{r\times r},\ J_{32} = \mathbf{0}_{r\times(n-r)},\ J_{33} = \mathbf{0}_{r\times r},\ J_{34} = \mathbf{0}_{r\times(n-r)}$$

and finally the sub-matrices in the fourth row are

$$J_{41} = \mathbf{0}_{(n-r)\times r},\ J_{42} = I_{(n-r)\times(n-r)},\ J_{43} = \mathbf{0}_{(n-r)\times r},\ J_{44} = \mathbf{0}_{(n-r)\times(n-r)}.$$

The eigenvalue equation can be written in vector form where the eigenvector is broken up to segments, $\boldsymbol{u},\ \boldsymbol{v},\ \bar{\boldsymbol{u}},\ \bar{\boldsymbol{v}}$ of sizes $r,\ n-r,\ r$ and $n-r$, respectively. Then we have

$$\begin{aligned} J_{11}\boldsymbol{u} \quad + J_{13}\bar{\boldsymbol{u}} + J_{14}\bar{\boldsymbol{v}} &= \lambda\boldsymbol{u} \\ J_{22}\boldsymbol{v} + J_{23}\bar{\boldsymbol{u}} + J_{24}\bar{\boldsymbol{v}} &= \lambda\boldsymbol{v} \\ \boldsymbol{u} \qquad\qquad &= \lambda\bar{\boldsymbol{u}} \\ \boldsymbol{v} \qquad\qquad &= \lambda\bar{\boldsymbol{v}}. \end{aligned}$$

Substituting the last two equations into the first two equations, we have

$$\left(J_{11}\lambda - \lambda^2 I + J_{13}\right)\bar{\boldsymbol{u}} + J_{14}\bar{\boldsymbol{v}} = \mathbf{0}$$

$$J_{23}\,\bar{\boldsymbol{u}} + \left(J_{22}\lambda - \lambda^2 I + J_{24}\right)\bar{\boldsymbol{v}} = \mathbf{0}$$

implying that the characteristic equation is the following:

$$\varphi(\lambda) = \det\begin{pmatrix} J_{11}\lambda - \lambda^2 I + J_{13} & J_{14} \\ J_{23} & J_{22}\lambda - \lambda^2 I + J_{24} \end{pmatrix} = 0 \tag{4.69}$$

which is a polynomial of degree $4n$.

Proposition 4.9 *The equilibrium with model* (4.68) *is locally asymptotically stable if the roots of equation* (4.69) *are inside the unit circle.*

Equation (4.69) is very complicated, so no general simple stability condition can be given.

The hyperbolic case can be examined in the same way, where the structure of the Jacobians is the same as in the linear case, but the elements of the blocks are different because of the different derivatives of the different best response functions.

4.2 Continuous Dynamics with Fixed Delays

In this section, we will revisit the models discussed earlier in this chapter and instead of discrete time scales, continuous time scales will be assumed. In the case when delays appear, in obtaining and implementing the firms' own information or those of the competitors, fixed delays will be assumed. The case of continuously distributed delays will be the subject of the next section.

Consider first the case of linear Cournot oligopolies, where the profit of each firm k is given in (4.4), and the best response of firm k is given by (4.31). Assume that the firms adjust their outputs in the direction toward their best responses, then their outputs are driven by the following dynamic rule:

$$\dot{q}_k(t) = K_k \left[R_k \left(\sum_{i \neq k}^{n} q_i(t) \right) - q_k(t) \right], \quad k = 1, 2, \ldots, n. \tag{4.70}$$

This equation can be derived from (4.32), if the output difference $q_k(t+1) - q_k(t)$ is replaced by the derivative $\dot{q}_k(t)$. This mechanism can be explained also as follows. If the best response of the firm is larger than the current output, then the firm wants to increase its output level. If the best response is below the current output level, then the firm wants to decrease it, and if they are equal, then the firm does not want any output change. Since the best response functions are linear, system (4.70) is linear, so local asymptotic stability implies global asymptotic stability.

The Jacobian of this system is

$$\boldsymbol{J} = \begin{pmatrix} -K_1 & -\frac{K_1\gamma}{2} & \cdots & -\frac{K_1\gamma}{2} \\ -\frac{K_2\gamma}{2} & -K_2 & \cdots & -\frac{K_2\gamma}{2} \\ \cdot & \cdot & \cdots & \cdot \\ -\frac{K_n\gamma}{2} & -\frac{K_n\gamma}{2} & \cdots & -K_n \end{pmatrix}.$$

The equilibrium with model (4.70) is asymptotically stable if and only if all eigenvalues are either real and negative, or have negative real parts. We can locate the eigenvalues of the Jacobian similarly to the discrete case. Notice that

$$\boldsymbol{J} = \boldsymbol{D} + \boldsymbol{a}\boldsymbol{b}^T,$$

where

$$\boldsymbol{D} = diag\left(-K_1\left(1-\frac{\gamma}{2}\right), \ldots, -K_n\left(1-\frac{\gamma}{2}\right)\right)$$

$$\boldsymbol{a} = \left(-\frac{K_1\gamma}{2}, \ldots, -\frac{K_n\gamma}{2}\right)^T$$

and

$$\boldsymbol{b}^T = (1, \ldots, 1).$$

So using again relation (4.34), the characteristic polynomial can be written as follows:

$$\begin{aligned}\varphi(\lambda) &= \det\left(\boldsymbol{D} + \boldsymbol{a}\boldsymbol{b}^T - \lambda\boldsymbol{I}\right),\\ &= \det(\boldsymbol{D} - \lambda\boldsymbol{I})\det\left(\boldsymbol{I} + (\boldsymbol{D} - \lambda\boldsymbol{I})^{-1}\boldsymbol{a}\boldsymbol{b}^T\right)\\ &= \prod_{k=1}^{n}\left[-K_k\left(1-\frac{\gamma}{2}\right) - \lambda\right]\left[1 + \sum_{k=1}^{n}\frac{\frac{K_k\gamma}{2}}{K_k\left(1-\frac{\gamma}{2}\right)+\lambda}\right].\end{aligned}$$

Assume that the firms are numbered such that the different K_k values satisfy $K_1 > K_2 > \cdots > K_S$ with multiplicities, $m_1, m_2, \ldots, m_S$. By adding the identical terms in the bracketed expression, we have

$$\varphi(\lambda) = (-1)^n \prod_{j=1}^{S}\left[K_j\left(1-\frac{\gamma}{2}\right)+\lambda\right]^{m_j}\left[1 + \sum_{j=1}^{S}\frac{\frac{T_j\gamma}{2}}{K_j\left(1-\frac{\gamma}{2}\right)+\lambda}\right] \tag{4.71}$$

where T_j is the sum of the identical speeds of adjustments. If $m_j = 1$, then $-K_j(1-\gamma/2)$ is not an eigenvalue, and if $m_j \geq 2$, then it is an eigenvalue. Since $0 < K_j \leq 1$ and $-1 \leq \gamma \leq +1$, these eigenvalues are real and negative. The other eigenvalues are the solutions of equation

$$g(\lambda) = 1 + \sum_{j=1}^{S}\frac{\frac{T_j\gamma}{2}}{K_j\left(1-\frac{\gamma}{2}\right)+\lambda} = 0. \tag{4.72}$$

The poles of $g(\lambda)$ are the $-K_j(1-\gamma/2)$ values, and all of them are negative. We next consider two cases.

If $\gamma > 0$, then the graph of $g(\lambda)$ is shown in Fig. 4.1, where all poles are negative. Therefore, all roots are negative, implying the asymptotical stability of the equilibrium.

If $\gamma < 0$, then the graph of $g(\lambda)$ is shown in Fig. 4.2, where all poles are again real and negative. All roots between the poles are all negative, and the largest solution above the largest pole is also negative if and only if $g(0) > 0$, which can be written as

$$1 + \sum_{k=1}^{n} \frac{\frac{K_k \gamma}{2}}{K_k \left(1 - \frac{\gamma}{2}\right)} = 1 + \sum_{k=1}^{n} \frac{\gamma}{2-\gamma} = 1 + \frac{n\gamma}{2-\gamma} > 0,$$

that is, the condition is the same as to guarantee positive equilibrium industry output,

$$\gamma > -\frac{2}{n-1}, \tag{4.73}$$

which holds for positive values of γ as well. We can summarize the above discussion as follows:

Proposition 4.10 *The equilibrium with model* (4.70) *is asymptotically stable if and only if* (4.73) *is satisfied.*

In comparison to the discrete case, notice that for negative values of γ the conditions are identical, however for $\gamma > 0$, the continuous time model always guarantees the asymptotical stability of the equilibrium.

Assume next that the firms have identical delay τ in the outputs of the rest of the industry. In this case, model (4.70) is modified as follows:

$$\dot{q}_k(t) = K_k \left[R_k \left(\sum_{i \neq k}^{n} q_i(t-\tau) \right) - q_k(t) \right], \quad k = 1, 2, \ldots, n, \tag{4.74}$$

or by using the actual form of the best responses

$$\dot{q}_k(t) = K_k \left(\frac{\alpha_k - c_k - \gamma \sum_{i \neq k}^{n} q_i(t-\tau)}{2} - q_k(t) \right).$$

The associated homogeneous equation is as follows:

$$\dot{q}_{k\varepsilon}(t) = K_k \left(\frac{-\gamma \sum_{i \neq k}^{n} q_{i\varepsilon}(t-\tau)}{2} - q_{k\varepsilon}(t) \right)$$

where $q_{k\varepsilon}(t)$ is the difference of $q_k(t)$ and its equilibrium level. The characteristic equation can be obtained by looking for the solutions in exponential forms, $q_{k\varepsilon}(t) = e^{\lambda t} u_k$, and substituting them into the homogeneous equation to have

$$\lambda e^{\lambda t} u_k = -\frac{K_k \gamma}{2} \sum_{i \neq k}^{n} e^{\lambda(t-\tau)} u_i - K_k e^{\lambda t} u_k$$

giving a system of homogeneous linear equations for the variables u_k

$$(\lambda + K_k)\, u_k + \frac{K_k \gamma}{2} \sum_{i \neq k}^{n} e^{-\lambda \tau} u_i = 0 \; (k = 1, 2, \ldots, n).$$

The determinant of this system provides the characteristic polynomial

$$\varphi(\lambda) = \det \begin{pmatrix} \lambda + K_1 & \frac{K_1 \gamma}{2} e^{-\lambda \tau} & \cdots & \frac{K_1 \gamma}{2} e^{-\lambda \tau} \\ \frac{K_2 \gamma}{2} e^{-\lambda \tau} & \lambda + K_2 & \cdots & \frac{K_2 \gamma}{2} e^{-\lambda \tau} \\ \cdot & \cdot & \cdots & \cdot \\ \frac{K_n \gamma}{2} e^{-\lambda \tau} & \frac{K_n \gamma}{2} e^{-\lambda \tau} & \cdots & \lambda + K_n \end{pmatrix}.$$

We can use again the same idea as in the discrete case by introducing the notation

$$\boldsymbol{D} = diag\left(\lambda + K_1 \left(1 - \frac{\gamma}{2} e^{-\lambda \tau}\right), \ldots, \lambda + K_n \left(1 - \frac{\gamma}{2} e^{-\lambda \tau}\right)\right),$$

$$\boldsymbol{a} = \left(\frac{K_1 \gamma}{2} e^{-\lambda \tau}, \ldots, \frac{K_n \gamma}{2} e^{-\lambda \tau}\right)^T$$

and

$$\boldsymbol{b}^T = (1, \ldots, 1)$$

to have

$$\begin{aligned} \varphi(\lambda) &= \det\left(\boldsymbol{D} + \boldsymbol{a}\boldsymbol{b}^T\right), \\ &= \det(\boldsymbol{D}) \det\left(\boldsymbol{I} + \boldsymbol{D}^{-1}\boldsymbol{a}\boldsymbol{b}^T\right) \\ &= \prod_{k=1}^{n} \left[\lambda + K_k \left(1 - \frac{\gamma}{2} e^{-\lambda \tau}\right)\right] \left[1 + \sum_{k=1}^{n} \frac{\frac{K_k \gamma}{2} e^{-\lambda \tau}}{\lambda + K_k \left(1 - \frac{\gamma}{2} e^{-\lambda \tau}\right)}\right]. \end{aligned} \tag{4.75}$$

Consider first equation

$$\lambda + K_k \left(1 - \frac{\gamma}{2} e^{-\lambda \tau}\right) = 0. \tag{4.76}$$

Without delay $\tau = 0$, so the solution is

$$\lambda = -K_k\left(1 - \frac{\gamma}{2}\right),$$

which is negative. As τ increases, at least one solution might change its real part from negative to positive leading to the instability of system (4.74). At the stability switch $\lambda = iv$ $(v > 0)$, which is then substituted into Eq. (4.76),

$$iv + K_k - \frac{K_k\gamma}{2}(\cos v\tau - i\sin v\tau) = 0.$$

Separating the real and imaginary parts,

$$\frac{K_k\gamma}{2}\cos v\tau = K_k \tag{4.77}$$

$$\frac{K_k\gamma}{2}\sin v\tau = -v \tag{4.78}$$

and by adding the squares of these equations

$$\frac{K_k^2\gamma^2}{4} = K_k^2 + v^2$$

implying that

$$v^2 = K_k^2\left(\frac{\gamma^2}{4} - 1\right) < 0.$$

Therefore, there is no stability switch, and all roots of equation (4.76) have negative real parts for all $\tau > 0$. The other eigenvalues are solutions of equation

$$1 + \sum_{k=1}^{n} \frac{\frac{K_k\gamma}{2}e^{-\lambda\tau}}{\lambda + K_k\left(1 - \frac{\gamma}{2}e^{-\lambda\tau}\right)} = 0. \tag{4.79}$$

This equation is complicated, and no simple conditions can be derived in general. Therefore, we consider the symmetric case when $K_1 = \cdots = K_n = K$. Then the equation can rewritten as

$$\lambda + K\left(1 - \frac{\gamma}{2}e^{-\lambda\tau}\right) + \frac{nK\gamma}{2}e^{-\lambda\tau} = \lambda + K + \frac{K(n-1)\gamma}{2}e^{-\lambda\tau} = 0. \tag{4.80}$$

Without delay the solution is

$$\lambda = -K\left(1 + \frac{n-1}{2}\gamma\right)$$

which is negative if $\gamma > -2/(n-1)$. In this case, we have stability without delay, as it was already established in condition (4.73). At a stability switch $\lambda = iv$ $(v > 0)$ and by substituting it into Eq. (4.80), we have

$$iv + K + \frac{K(n-1)\gamma}{2}(\cos v\tau - i \sin v\tau) = 0$$

or

$$\frac{K(n-1)\gamma}{2} \cos v\tau = -K, \tag{4.81}$$

$$\frac{K(n-1)\gamma}{2} \sin v\tau = v. \tag{4.82}$$

By adding the squares of these equations, we have

$$v^2 = K^2 \left(\frac{(n-1)^2\gamma^2}{4} - 1 \right).$$

Solution exists only if $(n-1)\,|\gamma| > 2$, in which case

$$v^* = K\sqrt{\frac{(n-1)^2\gamma^2}{4} - 1}.$$

Notice that for $n = 2$ and $n = 3$, the condition for existence of positive v cannot be satisfied, so there is no stability switch; therefore, duopolies and triopolies are asymptotically stable for all $\tau > 0$.

Assume next that $n > 3$. From Eq. (4.81), we have the critical value for τ as

$$\tau_n^* = \begin{cases} \dfrac{1}{v^*}\left[\cos^{-1}\left(\dfrac{-2}{(n-1)\gamma}\right) + 2n\pi\right] & \text{if } \gamma > 0, \\[2ex] \dfrac{1}{v^*}\left[2\pi - \cos^{-1}\left(\dfrac{-2}{(n-1)\gamma}\right) + 2n\pi\right] & \text{if } \gamma < 0. \end{cases} \tag{4.83}$$

The direction of stability switch is now determined by Hopf bifurcation. Consider τ as the bifurcation parameter and assume that the eigenvalue is a function of τ: $\lambda = \lambda(\tau)$. Implicitly differentiating equation (4.80) with respect to τ gives an equation for $\lambda'(\tau)$ as follows:

$$\lambda'(\tau) + \frac{K(n-1)\gamma}{2} e^{-\lambda(\tau)\tau}\left[-\lambda'(\tau)\tau - \lambda(\tau)\right] = 0$$

so

$$\lambda'(\tau) = \frac{\dfrac{K(n-1)\gamma}{2}\lambda(\tau)e^{-\lambda(\tau)\tau}}{1 - \dfrac{K(n-1)\gamma}{2}\tau e^{-\lambda(\tau)\tau}}.$$

From Eq. (4.80), we know that

$$\frac{K(n-1)\gamma}{2}e^{-\lambda(\tau)\tau} = -\lambda(\tau) - K,$$

so

$$\lambda'(\tau) = \frac{-\lambda(\tau)^2 - \lambda(\tau)K}{1 + \lambda(\tau)\tau + K\tau}.$$

At $\lambda = iv$,

$$\lambda'(\tau) = \frac{v^2 - ivK}{1 + K\tau + iv\tau}\frac{1 + K\tau - iv\tau}{1 + K\tau - iv\tau}.$$

Its real part has the same sign as

$$v^2(1 + K\tau) - v^2 K\tau = v^2 > 0$$

showing that at critical values, at least one pair of eigenvalues changes real part from negative to positive.

Proposition 4.11 *In the symmetric case, when $K_1 = \cdots = K_n = K$, the system is asymptotically stable without delay if equilibrium industry output is positive. Increasing the value of τ stability is lost at τ_0^*, and stability cannot be regained later. At $\tau = \tau_0^*$ Hopf bifurcation occurs giving the possibility of limit cycles.*

Assume next that there are two groups of firms. Firms $1, 2, \ldots, r$ have K as their common speed of adjustment, and firms $r+1, \ldots, n$ are with common speed of adjustment $\bar{K}$. Then Eq. (4.79) can be rewritten as

$$\left[\lambda + K\left(1 - \frac{\gamma}{2}e^{-\lambda\tau}\right)\right]\left[\lambda + \bar{K}\left(1 - \frac{\gamma}{2}e^{-\lambda\tau}\right)\right]$$
$$+\frac{rK\gamma}{2}e^{-\lambda\tau}\left[\lambda + \bar{K}\left(1 - \frac{\gamma}{2}e^{-\lambda\tau}\right)\right] + \frac{(n-r)\bar{K}\gamma}{2}e^{-\lambda\tau}\left[\lambda + K\left(1 - \frac{\gamma}{2}e^{-\lambda\tau}\right)\right] = 0$$

or by rearranging the terms

$$\left(\lambda^2 + \left(K + \bar{K}\right)\lambda + K\bar{K}\right) + \tfrac{\gamma}{2}\left[(n-2)K\bar{K} + \lambda\left((r-1)K + (n-r-1)\bar{K}\right)\right]$$
$$\times\, e^{-\lambda\tau} - \tfrac{(n-1)K\bar{K}\gamma^2}{4}e^{-2\lambda\tau} = 0.$$

This equation can be written as

$$1 + a_1(\lambda)e^{-\lambda\tau} + a_2(\lambda)e^{-2\lambda\tau} = 0 \tag{4.84}$$

where

$$a_1(\lambda) = \frac{\gamma\left[(n-2)K\bar{K} + \lambda\left((r-1)K + (n-r-1)\bar{K}\right)\right]}{2\left(\lambda^2 + \left(K + \bar{K}\right)\lambda + K\bar{K}\right)}$$

and

$$a_2(\lambda) = -\frac{(n-1)K\bar{K}\gamma^2}{4\left(\lambda^2 + \left(K+\bar{K}\right)\lambda + K\bar{K}\right)}.$$

This is the form of the two-delay equation (A7) introduced in Appendix A. So we can treat Eq. (4.84) as a two-delay equation and find the stability switching curves (τ_1, τ_2). The critical values of τ are the τ_1 values at the intersections of the stability switching curves and the $\tau_2 = 2\tau_1$ line.

The linear Bertrand oligopoly also has linear best response functions, so their analysis can be done in the same way as it was shown before. The only difference is that γ has to be replaced by $\bar{\gamma}$ as it was introduced for the discrete equation (4.56). Mixed oligopolies with fixed delays can be examined similarly to the discrete case with very similar block matrices. In the no-delay case, the Jacobian is obtained by subtracting the identity matrix from that of the discrete case. If delay is introduced, then we have similar matrix structures with exponential factors in the appropriate matrix elements. The entire analysis shown for the earlier cases remains valid with minor changes.

We can now turn our attention to hyperbolic oligopolies when the continuous time counterpart of model (4.46) has the form

$$\dot{q}_k(t) = K_k\left(\sqrt{\frac{\sum_{i\neq k}^n q_i(t)}{c_k}} - \sum_{i\neq k}^n q_i(t) - q_k(t)\right) \text{ for } k = 1, 2, \ldots, n. \quad (4.85)$$

The Jacobian matrix of this system is obtained by subtracting the identity matrix from the Jacobian of system (4.46):

$$\boldsymbol{J} = \begin{pmatrix} -K_1 & K_1\gamma_1 & \cdots & K_1\gamma_1 \\ K_2\gamma_2 & -K_2 & \cdots & K_2\gamma_2 \\ \cdot & \cdot & \cdots & \cdot \\ K_n\gamma_n & K_n\gamma_n & \cdots & -K_n \end{pmatrix}$$

where γ_k is the derivative of the best response function of firm k at the equilibrium. We have already proved in the previous section that

$$\gamma_k = \frac{C}{2c_k(n-1)} - 1$$

with $C = \sum_{i=1}^n c_i$. Similarly to the previously discussed cases, we can derive a simple formula of the characteristic equation of matrix $\boldsymbol{J}$. For this purpose, introduce the following notation:

$$\boldsymbol{D} = diag\left(-K_1\left(1+\gamma_1\right), \ldots, -K_n\left(1+\gamma_n\right)\right)$$

$$\boldsymbol{a} = \left(K_1\gamma_1, \ldots, K_n\gamma_n\right)^T$$

and

$$\boldsymbol{b}^T = (1, \ldots, 1).$$

Then it is easy to see based on relation (4.34) that

$$\begin{aligned}\varphi(\lambda) &= \det\left(\boldsymbol{D} + \boldsymbol{a}\boldsymbol{b}^T - \lambda \boldsymbol{I}\right), \\ &= \det(\boldsymbol{D} - \lambda\boldsymbol{I}) \det\left(\boldsymbol{I} + (\boldsymbol{D} - \lambda\boldsymbol{I})^{-1} \boldsymbol{a}\boldsymbol{b}^T\right) \\ &= \prod_{k=1}^{n} [-K_k(1+\gamma_k) - \lambda] \left[1 - \sum_{k=1}^{n} \frac{K_k \gamma_k}{K_k(1+\gamma_k) + \lambda}\right].\end{aligned}$$

From the first factor

$$\lambda = -K_k(1+\gamma_k)$$

which is negative if $\gamma_k > -1$, which holds, since C, c_k and $(n-1)$ are all positive. The other eigenvalues are the solutions of equation

$$g(\lambda) = 1 - \sum_{k=1}^{n} \frac{K_k \gamma_k}{K_k(1+\gamma_k) + \lambda} = 0. \tag{4.86}$$

Since the sign of γ_k is indeterminate, no general simple stability condition can be given.

Proposition 4.12 *The equilibrium with model* (4.85) *is locally asymptotically stable if all solutions of equation* (4.86) *are negative or have negative real parts.*

Consider now the special case, where $c_1 = \cdots = c_n = c$ and $n \geq 3$. Then

$$\gamma_1 = \cdots = \gamma_n = \gamma = \frac{2-n}{2(n-1)}$$

which is between $-1/2$ and $-1/4$. So if the marginal costs are sufficiently close to each other, then all γ_k values are negative and between -1 and 0. So all poles of $g(\lambda)$ are negative. Notice that $g'(\lambda) < 0$ and the graph of $g(\lambda)$ is similar to that shown in Fig. 4.1. The roots between the poles are negative and the last root below the smallest pole is also negative, implying that the equilibrium is locally asymptotically stable. If $n = 2$, then $\boldsymbol{J}$ is diagonal with negative eigenvalues, so the equilibrium is asymptotically stable.

Assume next that there is a delay for the firms in obtaining the values of the output of the rest of the industry. Then the dynamic equation (4.85) is modified as

$$\dot{q}_k(t) = K_k \left(\sqrt{\frac{\sum_{i \neq k}^{n} q_i(t-\tau)}{c_k}} - \sum_{i \neq k}^{n} q_i(t-\tau) - q_k(t) \right) \text{ for } k = 1, 2, \dots, n \tag{4.87}$$

where $\tau > 0$ is the delay. Since

$$\frac{\partial \dot{q}_k(t)}{\partial q_k(t)} = -K_k \text{ and } \frac{\partial \dot{q}_k(t)}{\partial q_i(t-\tau)} = K_k \gamma_k,$$

the linearized equation has the form

$$\dot{q}_{k\varepsilon}(t) = -K_k q_{k\varepsilon}(t) + K_k \gamma_k \sum_{i \neq k}^{n} q_{i\varepsilon}(t-\tau) \text{ for } k = 1, 2, \dots, n.$$

As usual, we are looking for the solution in exponential form

$$q_{k\varepsilon}(t) = e^{\lambda t} u_k$$

and by substituting them into this linearized equation, we have

$$\lambda e^{\lambda t} u_k = -K_k e^{\lambda t} u_k + K_k \gamma_k \sum_{i \neq k}^{n} e^{\lambda(t-\tau)} u_i$$

or

$$(\lambda + K_k) u_k - K_k \gamma_k \sum_{i \neq k}^{n} e^{-\lambda\tau} u_i = 0.$$

The determinant of the coefficient matrix is zero, so the characteristic equation can be written as

$$\varphi(\lambda) = \det \begin{pmatrix} \lambda + K_1 & -K_1\gamma_1 e^{-\lambda\tau} & \cdots & -K_1\gamma_1 e^{-\lambda\tau} \\ -K_2\gamma_2 e^{-\lambda\tau} & \lambda + K_2 & \cdots & -K_2\gamma_2 e^{-\lambda\tau} \\ \cdot & \cdot & \cdots & \cdot \\ -K_n\gamma_n e^{-\lambda\tau} & -K_n\gamma_n e^{-\lambda\tau} & \cdots & \lambda + K_n \end{pmatrix} = 0.$$

Introduce notation

$$\boldsymbol{D} = diag\left(\lambda + K_1\left(1 + \gamma_1 e^{-\lambda\tau}\right), \dots, \lambda + K_n\left(1 + \gamma_n e^{-\lambda\tau}\right)\right)$$

$$\boldsymbol{a} = \left(-K_1\gamma_1 e^{-\lambda\tau}, \dots, -K_n\gamma_n e^{-\lambda\tau}\right)^T$$

and

$$\boldsymbol{b}^T = (1, \dots, 1)$$

to have

$$\begin{aligned}\varphi(\lambda) &= \det\left(\boldsymbol{D} + \boldsymbol{a}\boldsymbol{b}^T\right), \\ &= \det(\boldsymbol{D}) \det\left(\boldsymbol{I} + \boldsymbol{D}^{-1}\boldsymbol{a}\boldsymbol{b}^T\right) \\ &= \prod_{k=1}^{n}\left[\lambda + K_k\left(1 + \gamma_k e^{-\lambda\tau}\right)\right]\left[1 - \sum_{k=1}^{n} \frac{K_k \gamma_k e^{-\lambda\tau}}{K_k\left(1 + \gamma_k e^{-\lambda\tau}\right) + \lambda}\right].\end{aligned} \tag{4.88}$$

The first equation is the same as (4.76) examined earlier. Without delay, the root is $-K_k(1+\gamma_k)$, and since $\gamma_k > -1$, this root is negative. Increasing the value of τ from zero, stability might be lost, when $\lambda = iv$ with $v > 0$. Since γ_k takes the place of $-\gamma/2$, by repeating the derivation with Eq. (4.76) we can conclude that at the critical point

$$v^2 = K_k^2\left(\gamma_k^2 - 1\right).$$

Therefore, there is no stability switch if $|\gamma_k| \le 1$. We have seen already that

$$\gamma_k = \frac{C}{2c_k(n-1)} - 1 > -1$$

and

$$\gamma_k \le 1 \text{ if } \sum_{i=1}^{n} c_i \le 4c_k(n-1) \text{ or } c_k \ge \frac{n}{4(n-1)}\left(\frac{1}{n}\sum_{i=1}^{n} c_i\right)$$

which is the case if there is no firm with too low marginal cost.

Similarly to the previous symmetric case, we assume that $K_1 = \cdots = K_n = K$ and $c_1 = \cdots = c_k = c$ so $\gamma_1 = \cdots = \gamma_n = \gamma$ where

$$\gamma = \frac{2-n}{2(n-1)} \text{ and } |\gamma| \le 1.$$

The other eigenvalues are solutions of equation

$$\lambda + K - (n-1)K\gamma e^{-\lambda\tau} = 0$$

which is identical to Eq. (4.80) with the only difference that $\gamma/2$ is now replaced with $-\gamma$. Therefore, Proposition 4.11 remains valid with the only difference that in this case

$$\tau_0^* = \begin{cases} \dfrac{1}{v^*}\left[2\pi - \cos^{-1}\left(\dfrac{1}{(n-1)\gamma}\right)\right] & \text{if } \gamma > 0, \\[2ex] \dfrac{1}{v^*}\left[\cos^{-1}\left(\dfrac{1}{(n-1)\gamma}\right)\right] & \text{if } \gamma < 0 \end{cases}$$

where

$$v^* = K\sqrt{(n-1)^2\gamma^2 - 1}.$$

Notice that stability switch occurs if $(n-1)^2\gamma^2 - 1 > 0$, which can be rewritten as

$$\left(\frac{2-n}{2}\right)^2 > 1$$

which occurs if $n > 4$. That is, in the symmetric case, we need at least five firms to have instability, since without delay the equilibrium is stable for all $n \geq 2$.

Consider next hyperbolic Bertrand oligopolies. The best responses of the firms are given in the previous section by Eq. (4.57). Let γ_{kj} denote the derivative of $R_k\left(\boldsymbol{p}_{-k}\right)$ with respect to p_j. From the best response function (4.57) for $j \neq k$,

$$\begin{aligned}\gamma_{kj} &= \frac{1}{2\sqrt{\frac{c_k b_{kk}^{-1}}{\sum_{j\neq k}^{n} b_{kj}^{-1}\frac{1}{p_j}}}}\frac{-c_k b_{kk}^{-1}\frac{b_{kj}^{-1}}{p_j^2}}{\left(\sum_{j\neq k}^{n} b_{kj}^{-1}\frac{1}{p_j}\right)^2} \\ &= -\frac{b_{kj}^{-1}}{2p_j^2}\sqrt{\frac{c_k b_{kk}^{-1}}{\left(\sum_{j\neq k}^{n} b_{kj}^{-1}\frac{1}{p_j}\right)^3}}.\end{aligned} \tag{4.89}$$

Since the dynamic system is written as

$$\dot{p}_k(t) = K_k\left[R_k\left(\boldsymbol{p}_{-k}(t)\right) - p_k(t)\right], \tag{4.90}$$

the Jacobian has the form

$$\boldsymbol{J} = \begin{pmatrix} -K_1 & K_1\gamma_{12} & \cdots & K_1\gamma_{1n} \\ K_2\gamma_{21} & -K_2 & \cdots & K_2\gamma_{2n} \\ \cdot & \cdot & \cdots & \cdot \\ K_n\gamma_{n1} & K_n\gamma_{n2} & \cdots & -K_n \end{pmatrix}.$$

In this general form, no general result can be given.

Proposition 4.13 *The equilibrium with model* (4.90) *is locally asymptotically stable if all eigenvalues of* $\boldsymbol{J}$ *are real negative or complex with negative real parts.*

In the special symmetric case, where $K_1 = \cdots = K_n = K,\ c_1 = \cdots = c_n = c$ and $\Theta_{kj} = \Theta$ for all k and j, we have already showed with model (4.58) that

$$\gamma_{kj} = \frac{1}{2(n-1)} \text{ for all } k \text{ and } j.$$

So the Jacobian is the same as that of system (4.70) with $-\gamma/2$ is replaced by $\bar{\gamma} = 1/[2(n-1)] > 0$. From Proposition 4.10, we can see that in this case the equilibrium is locally asymptotically stable, since

$$-2\bar{\gamma} = -\frac{1}{n-1} > -\frac{2}{n-1}.$$

Assume next that there is a delay in obtaining information about the prices of the competitors. Then similarly to model (4.90), we have the following dynamic system:

$$\dot{p}_k(t) = K_k\left[R_k\left(\boldsymbol{p}_{-k}(t-\tau)\right) - p_k(t)\right]. \tag{4.91}$$

Linearization gives the linear system

$$\dot{p}_k(t) = -K_k p_k(t) + K_k \sum_{i\neq k}^{n} \gamma_{ki} p_i(t-\tau) \tag{4.92}$$

where γ_{ki} is the derivative of $R_k\left(\boldsymbol{p}_{-k}\right)$ with respect to p_i given by (4.89). As usual, the characteristic polynomial is obtained by substituting exponential solution forms

$$p_k(t) = e^{\lambda t} u_k$$

into Eq. (4.92) to have

$$\lambda e^{\lambda t} u_k = -K_k e^{\lambda t} u_k + K_k \sum_{i\neq k}^{n} \gamma_{ki} e^{\lambda(t-\tau)} u_i$$

and after simplification,

$$(\lambda + K_k)\, u_k - K_k \sum_{i\neq k}^{n} \gamma_{ki} e^{-\lambda\tau} u_i \text{ for } k = 1, 2, \ldots, n.$$

The determinant of this system is zero leading to the characteristic equation,

$$\varphi(\lambda) = \det\begin{pmatrix} \lambda + K_1 & -K_1\gamma_{12}e^{-\lambda\tau} & \cdots & -K_1\gamma_{1n}e^{-\lambda\tau} \\ -K_2\gamma_{21}e^{-\lambda\tau} & \lambda + K_2 & \cdots & -K_2\gamma_{2n}e^{-\lambda\tau} \\ \cdot & \cdot & \cdots & \cdot \\ -K_n\gamma_{n1}e^{-\lambda\tau} & -K_n\gamma_{n2}e^{-\lambda\tau} & \cdots & \lambda + K_n \end{pmatrix} = 0. \tag{4.93}$$

Proposition 4.14 *The equilibrium with model* (4.91) *is locally asymptotically stable if all eigenvalues are either real negative or complex with negative real parts.*

Since we cannot give simple stability conditions in general, for the sake of simplicity, consider the symmetric case, when $K_1 = \cdots = K_n = K$, $c_1 = \cdots = c_n = c$ and $\Theta_{kj} = \Theta$ for all k and j. We have seen in examining the model without delay that

$$\gamma_{kj} = \bar{\gamma} = \frac{1}{2(n-1)} \text{ for all } k \text{ and } j.$$

In this special case, the characteristic polynomial is the same as that of the linear model (4.74) with the only difference that in this case $\gamma/2$ is replaced by $-\bar{\gamma}$. So in derivation of Proposition 4.11, we have to use

$$\gamma = -2\bar{\gamma} = -\frac{1}{n-1}.$$

In examining equation (4.76), there is no stability switch, since $\gamma^2/4 - 1 < 0$. In Eq. (4.80), there is no stability switch either, since

$$\frac{(n-1)^2\gamma^2}{4} - 1 = -\frac{3}{4} < 0.$$

Therefore in the symmetric case, the equilibrium with model (4.91) is locally asymptotically stable with all $\tau > 0$.

Mixed oligopolies can be discussed similarly to the discrete time cases. The Jacobians in both cases are obtained from the Jacobians of the discrete cases by subtracting the identity matrix from them. The block structures of the Jacobians remain the same with slightly different matrix elements. Since no general results could be given in the discrete cases, the same holds for the continuous time models as well. The general matrix formulations were presented for the discrete linear case in the previous section, so they are not repeated here.

4.3 Continuous Dynamics with Distributed Delays

In this section, continuous time scales are examined similarly to the previous section; however, instead of fixed delays, continuously distributed delays are considered.

We start the discussion with linear Cournot oligopolies, where the profit of firm k is given as

$$\pi_k(q_1, \ldots, q_n) = q_k \left(\alpha_k - q_k - \gamma \sum_{i \neq k}^{n} q_i - c_k \right) \tag{4.94}$$

with $\alpha_k > c_k > 0$ and $-1 \leq \gamma \leq 1$. The best response of this firm is the following:

$$R_k(Q_k) = \frac{\alpha_k - c_k - \gamma Q_k}{2}$$

with $Q_k = \sum_{i \neq k}^{n} q_i$ being the output of the rest of the industry. In the previous section, dynamics with adjustments toward best responses was considered. Without delays, the equilibrium is asymptotically stable if and only if $\gamma > -2/(n-1)$. If each firm has continuously distributed delay in obtaining and implementing information on the output of the rest of the industry, then the dynamic equations have the general form:

$$\dot{q}_k(t) = K_k \left(\frac{\alpha_k - c_k - \gamma \sum_{i \neq k}^{n} q_i^e(t)}{2} - q_k(t) \right) \text{ for } k = 1, 2, \ldots, n \tag{4.95}$$

with

$$q_i^e(t) = \int_0^t W(t-s, m_i, T_i)\, q_i(s) ds \tag{4.96}$$

where the weighting function is the same as in the previous chapters.

If $q_{k\varepsilon}(t)$ and $q_{i\varepsilon}^e(t)$ denote the differences of $q_k(t)$ and $q_i^e(t)$ from their equilibrium levels, then from (4.95), we get a homogeneous equation,

$$\dot{q}_{k\varepsilon}(t) = K_k \left[-\frac{\gamma}{2} \sum_{i \neq k}^{n} \int_0^t W(t-s, m_i, T_i)\, q_{i\varepsilon}(s) ds - q_{k\varepsilon}(t) \right]. \tag{4.97}$$

By considering exponential solutions

$$q_{k\varepsilon}(t) = e^{\lambda t} u_k \text{ for } k = 1, 2, \ldots, n$$

and letting $t \to \infty$, based on Eq. (B5) of Appendix B, we have the characteristic equations

$$(\lambda + K_k)\, u_k + \frac{K_k \gamma}{2} \sum_{i \neq k}^{n} \left(1 + \frac{T_i \lambda}{\bar{m}_i}\right)^{-(m_i+1)} u_i = 0 \text{ for } k = 1, 2, \ldots, n$$

where

$$\bar{m}_i = \begin{cases} 1 & \text{if } m_i = 0, \\ m_i & \text{if } m_i \geq 1. \end{cases}$$

The determinant of these linear equations for unknowns u_k $(1 \leq k \leq n)$ is zero which provides the characteristic function:

$$\varphi(\lambda) = \det \begin{pmatrix} \lambda + K_1 & \frac{K_1\gamma}{2}\left(1+\frac{T_2\lambda}{\bar{m}_2}\right)^{-(m_2+1)} & \cdots & \frac{K_1\gamma}{2}\left(1+\frac{T_n\lambda}{\bar{m}_n}\right)^{-(m_n+1)} \\ \frac{K_2\gamma}{2}\left(1+\frac{T_1\lambda}{\bar{m}_1}\right)^{-(m_1+1)} & \lambda + K_2 & \cdots & \frac{K_2\gamma}{2}\left(1+\frac{T_n\lambda}{\bar{m}_n}\right)^{-(m_n+1)} \\ \cdot & \cdot & \cdots & \cdot \\ \frac{K_n\gamma}{2}\left(1+\frac{T_1\lambda}{\bar{m}_1}\right)^{-(m_1+1)} & \frac{K_n\gamma}{2}\left(1+\frac{T_2\lambda}{\bar{m}_2}\right)^{-(m_2+1)} & \cdots & \lambda + K_n \end{pmatrix}. \tag{4.98}$$

Introduce the following matrix and vectors:

$$\boldsymbol{D} = diag\left(K_1 + \lambda - \frac{K_1\gamma}{2}\left(1+\frac{T_1\lambda}{\bar{m}_1}\right)^{-(m_1+1)}, \ldots, K_n + \lambda - \frac{K_n\gamma}{2}\left(1+\frac{T_n\lambda}{\bar{m}_n}\right)^{-(m_n+1)}\right)$$

$$\boldsymbol{a} = \left(\frac{K_1\gamma}{2}, \ldots, \frac{K_n\gamma}{2}\right)^T$$

and

$$\boldsymbol{b}^T = \left(\left(1+\frac{T_1\lambda}{\bar{m}_1}\right)^{-(m_1+1)}, \ldots, \left(1+\frac{T_n\lambda}{\bar{m}_n}\right)^{-(m_n+1)}\right)$$

to have

$$\begin{aligned} \varphi(\lambda) &= \det\left(\boldsymbol{D} + \boldsymbol{a}\boldsymbol{b}^T\right), \\ &= \det(\boldsymbol{D}) \det\left(\boldsymbol{I} + \boldsymbol{D}^{-1}\boldsymbol{a}\boldsymbol{b}^T\right) \\ &= \prod_{k=1}^{n}\left[K_k + \lambda - \frac{K_k\gamma}{2}\left(1+\frac{T_k\lambda}{\bar{m}_k}\right)^{-(m_k+1)}\right]\left[1 + \sum_{k=1}^{n} \frac{\frac{K_k\gamma}{2}\left(1+\frac{T_k\lambda}{\bar{m}_k}\right)^{-(m_k+1)}}{K_k+\lambda-\frac{K_k\gamma}{2}\left(1+\frac{T_k\lambda}{\bar{m}_k}\right)^{-(m_k+1)}}\right]. \end{aligned}$$

Consider first equation

$$K_k + \lambda - \frac{K_k\gamma}{2}\left(1+\frac{T_k\lambda}{\bar{m}_k}\right)^{-(m_k+1)} = 0$$

which can be rewritten as

$$(K_k + \lambda)\left(1+\frac{T_k\lambda}{\bar{m}_k}\right)^{m_k+1} - \frac{K_k\gamma}{2} = 0. \tag{4.99}$$

This is a polynomial equation of degree $m_k + 2$. We can easily prove that all roots are real negative or complex with negative real parts. Assume in contrary that λ is a root with $\text{Re}\lambda \geq 0$. Then

$$|K_k + \lambda| \geq K_k, \quad \left|1 + \frac{T_k \lambda}{\bar{m}_k}\right| \geq 1 \text{ but } \left|\frac{K_k \gamma}{2}\right| < |K_k|$$

so λ cannot satisfy Eq. (4.99). The other eigenvalues solve equation

$$1 + \sum_{k=1}^{n} \frac{\frac{K_k \gamma}{2}}{(K_k + \lambda)\left(1 + \frac{T_k \lambda}{\bar{m}_k}\right)^{m_k+1} - \frac{K_k \gamma}{2}} = 0. \tag{4.100}$$

Proposition 4.15 *The equilibrium with model* (4.95) *is asymptotically stable if and only if all roots of equation* (4.100) *are either real negative or complex with negative real parts.*

In the general form, it is impossible to give simple stability conditions. Therefore, we consider the symmetric case

$$K_1 = \cdots = K_n = K, \; m_1 = \cdots = m_n = m \text{ and } T_1 = \cdots = T_n = T.$$

Then Eq. (4.100) can be rewritten as

$$(K + \lambda)\left(1 + \frac{T\lambda}{\bar{m}}\right)^{m+1} + \frac{(n-1)K}{2}\gamma = 0. \tag{4.101}$$

Assume that for a solution, $\text{Re}\lambda \geq 0$. Then

$$|K + \lambda| \geq K \text{ and } \left|1 + \frac{T\lambda}{\bar{m}}\right| \geq 1,$$

so λ cannot be a solution if

$$\frac{(n-1)|\gamma|}{2} < 1,$$

that is

$$-\frac{2}{n-1} < \gamma < \frac{2}{n-1}.$$

It is very complicated to examine the roots of the general equation (4.101) if this condition does not hold, so we consider special cases. Without delay $T = 0$ and the equilibrium is asymptotically stable if $\gamma > -2/(n-1)$ as shown before.

If $m = 0$, then the equation is quadratic:

$$(K + \lambda)(1 + T\lambda) + \frac{(n-1)K}{2}\gamma = T\lambda^2 + (1 + TK)\lambda + K\left(1 + \frac{n-1}{2}\gamma\right) = 0.$$

The equilibrium is asymptotically stable if and only if all coefficients are positive, so the stability condition is the same which guarantees positive equilibrium industry output or stability without delay:

$$\gamma > -\frac{2}{n-1}. \tag{4.102}$$

Assume next that $m = 1$, when Eq. (4.101) is cubic,

$$(K+\lambda)\left(1+2T\lambda+T^2\lambda^2\right)+\frac{(n-1)K}{2}\gamma = a_0\lambda^3+a_1\lambda^2+a_2\lambda+a_3 = 0$$

with

$$a_0 = T^2,\ a_1 = KT^2+2T,\ a_2 = 1+2KT \text{ and } a_3 = K\left(1+\frac{n-1}{2}\gamma\right).$$

The stability criteria are $a_3 > 0$ and $a_1a_2 > a_0a_3$ as shown in Appendix B. Notice that $a_3 > 0$ if (4.102) holds and $a_1a_2 > a_0a_3$ if

$$\left(KT^2+2T\right)(1+2KT) > KT^2\left(1+\frac{n-1}{2}\gamma\right)$$

which is a quadratic inequality in KT:

$$2\,(KT)^2+\left(4-\frac{n-1}{2}\gamma\right)KT+2 > 0. \tag{4.103}$$

The left hand side is a convex parabola with roots

$$KT_{1,2} = \frac{1}{4}\left[\frac{n-1}{2}\gamma-4\pm\sqrt{\left(\frac{n-1}{2}\gamma\right)^2-4(n-1)\gamma}\right]. \tag{4.104}$$

The discriminant is negative if

$$0 < \gamma < \frac{16}{n-1},$$

so with $\gamma > 0$ and $n \le 16$, there is no real root implying that (4.103) holds for all $TK > 0$ making the equilibrium asymptotically stable. If $\gamma = 0$, then $KT_{1,2} = -1$, so (4.103) holds for all positive values of KT. If $\gamma < 0$, then both roots are negative, so the equilibrium is asymptotically stable. If $\gamma > 16/(n-1)$, then both roots are positive and (4.103) holds if

$$KT < KT_1 \text{ or } KT > KT_2$$

Fig. 4.4 Stability region for the symmetric model (4.95)

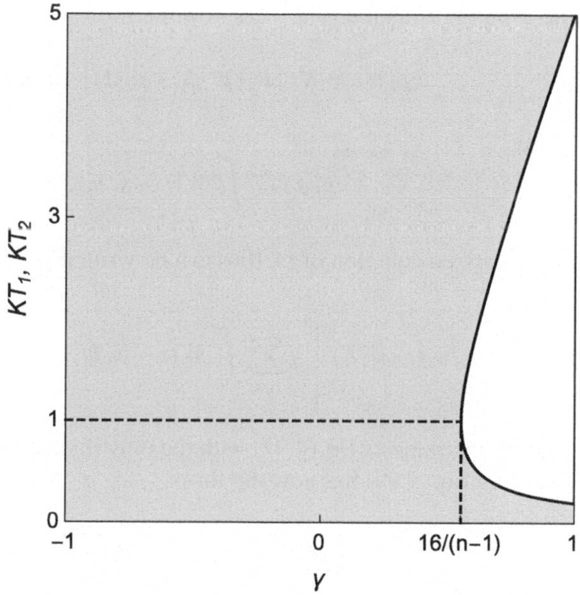

where KT_1 is the smaller root and KT_2 is the larger root. Figure 4.4 shows KT_1 and KT_2 as function of γ. The shaded area is the stability region. At $\gamma = 1$,

$$KT_{1,2} = \frac{n - 9 \pm \sqrt{n^2 - 18n + 17}}{8}.$$

Proposition 4.16 *In the symmetric case, the equilibrium with model* (4.95) *is asymptotically stable if*

$$\begin{aligned}
&(a)\ \textit{there is no delay or } m = 0 \textit{ with } \gamma > -\frac{2}{n-1} \\
&(b)\ m = 1 \textit{ if either } \gamma < \frac{16}{n-1} \\
&\qquad \textit{or } \gamma = \frac{16}{n-1} \textit{ and } KT \neq 1 \\
&\qquad \textit{or } \frac{16}{n-1} < \gamma \leq 1 \textit{ with } KT < KT_1 \textit{ or } KT > KT_2
\end{aligned}$$

where $KT_{1,2}$ are given in (4.104).

Consider next the linear Bertrand oligopoly case. The best responses of the firms are given in Sect. 4.1 by Eq. (4.55) where

$$\frac{\partial R_k(\boldsymbol{p}_{-k})}{\partial p_i} = \frac{\gamma}{2\left[1 + (n-2)\gamma\right]} = \bar{\gamma} \text{ for } i \neq k. \tag{4.105}$$

The dynamic equation (4.95) has now the form

$$\dot{p}_k(t) = K_k \left[R_k(\mathbf{p}^e_{-k}) - p_k(t) \right] \text{ for } k = 1, 2, \ldots, n \tag{4.106}$$

where

$$p_i^e(t) = \int_0^t W(t-s, m_i, T_i) p_i(s) ds.$$

The linearized equation of (4.106) can be written as

$$\dot{p}_{k\varepsilon}(t) = K_k \left[\bar{\gamma} \sum_{i \neq k}^{n} \int_0^t W(t-s, m_i, T_i) p_{i\varepsilon}(s) ds - p_{k\varepsilon}(t) \right]$$

which is the same as Eq. (4.97) with the only difference that $-\gamma/2$ is replaced by $\bar{\gamma}$. Therefore Eq. (4.99) has now the form

$$(K_k + \lambda)\left(1 + \frac{T_k \lambda}{\bar{m}_k}\right)^{m_k+1} + K_k \bar{\gamma} = 0. \tag{4.107}$$

Similarly to Eq. (4.99) it is easy to see that if $|\bar{\gamma}| < 1$, then all roots are real negative or have negative real parts. Equation (4.100) is now the following:

$$1 - \sum_{k=1}^{n} \frac{K_k \bar{\gamma}}{(K_k + \lambda)\left(1 + \frac{T_k \lambda}{\bar{m}_k}\right)^{m_k+1} + K_k \bar{\gamma}} = 0. \tag{4.108}$$

Proposition 4.17 *The equilibrium with model* (4.106) *is asymptotically stable if all roots of equations* (4.107) *and* (4.108) *are real negative or with negative real parts.*

In the symmetric case, these equations become

$$(K + \lambda)\left(1 + \frac{T\lambda}{\bar{m}}\right)^{m+1} + K\bar{\gamma} = 0 \tag{4.109}$$

and

$$(K + \lambda)\left(1 + \frac{T\lambda}{\bar{m}}\right)^{m+1} - (n-1)K\bar{\gamma} = 0. \tag{4.110}$$

If $T = 0$, then the solutions are negative if

$$-1 < \bar{\gamma} < \frac{1}{n-1}.$$

If $m = 0$, then these equations are quadratic,

$$(K+\lambda)(1+T\lambda)+K\bar{\gamma}=T\lambda^2+(1+TK)\lambda+K(1+\bar{\gamma})=0$$

and

$$(K+\lambda)(1+T\lambda)-(n-1)K\bar{\gamma}=T\lambda^2+(1+TK)\lambda+K(1-(n-1)\bar{\gamma})=0.$$

The equilibrium is asymptotically stable if all coefficients are positive. The condition is identical to the no-delay case:

$$-1<\bar{\gamma}<\frac{1}{n-1}. \tag{4.111}$$

Assume next that $m=1$, then Eqs. (4.109) and (4.110) are cubic:

$$(K+\lambda)\left(1+2T\lambda+T^2\lambda^2\right)+K\bar{\gamma}=a_0\lambda^3+a_1\lambda^2+a_2\lambda+a_3=0$$

with

$$a_0=T^2,\ a_1=KT^2+2T,\ a_2=1+2KT \text{ and } a_3=K(1+\bar{\gamma})$$

and

$$(K+\lambda)\left(1+2T\lambda+T^2\lambda^2\right)-(n-1)K\bar{\gamma}=a_0\lambda^3+a_1\lambda^2+a_2\lambda+a_3=0$$

with

$$a_0=T^2,\ a_1=KT^2+2T,\ a_2=1+2KT \text{ and } a_3=K(1-(n-1)\bar{\gamma}).$$

All coefficients are positive if (4.111) holds. Then the stability conditions are based on Theorem B.1 of Appendix B as follows:

$$\left(KT^2+2T\right)(1+2KT)>KT^2(1+\bar{\gamma}) \tag{4.112}$$

and

$$\left(KT^2+2T\right)(1+2KT)>KT^2(1-(n-1)\bar{\gamma}). \tag{4.113}$$

If $\bar{\gamma}>0$, then the first inequality is the stronger, if $\bar{\gamma}=0$, then they are identical and if $\bar{\gamma}<0$, then the second inequality is the stronger. We have already seen that the concavity of the profit functions requires that

$$\gamma<-\frac{1}{n-2} \text{ or } \gamma>-\frac{1}{n-1} \text{ so } \bar{\gamma}<1.$$

Assume first that $0\le\bar{\gamma}\le 1$. Then (4.112) is a quadratic inequality in KT:

$$2(KT)^2+(4-\bar{\gamma})KT+2>0$$

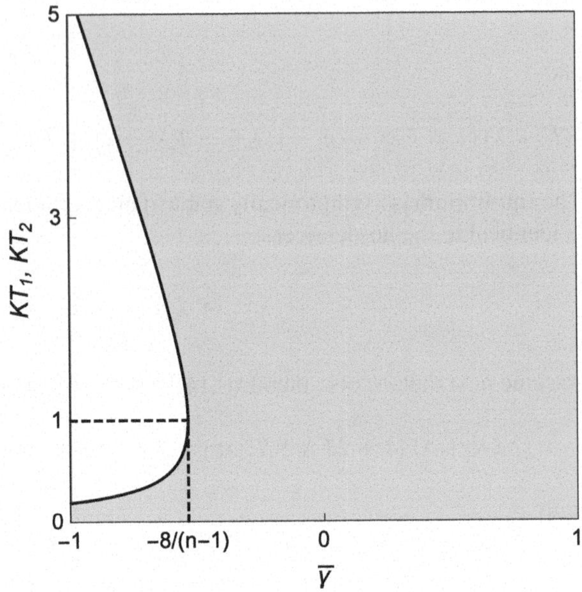

Fig. 4.5 Stability region for the symmetric model (4.106)

where all coefficients are positive, so the inequality holds for all $KT > 0$ implying the asymptotical stability of the equilibrium.

Assume next that $\bar{\gamma} < 0$. Then (4.113) can be rewritten as

$$2\,(KT)^2 + (4 + (n-1)\bar{\gamma})KT + 2 > 0.$$

The discriminant of the left hand side is $(n-1)^2\bar{\gamma}^2 + 8(n-1)\bar{\gamma}$, which is negative if

$$-\frac{8}{n-1} < \bar{\gamma} < 0.$$

Otherwise, the roots of the left hand side are

$$KT_{1,2} = \frac{1}{4}\left[-4 - (n-1)\bar{\gamma} \pm \sqrt{(n-1)^2\bar{\gamma}^2 + 8(n-1)\bar{\gamma}}\right]. \tag{4.114}$$

At $\bar{\gamma} = 0$, $KT_{1,2} = -1$, so (4.114) holds for all $KT > 0$, therefore the equilibrium is asymptotically stable. If $\bar{\gamma} = -\frac{8}{n-1}$, then $KT_1 = KT_2 = 1$, so the equilibrium is asymptotically stable if $KT \neq 1$. If $\bar{\gamma} < -\frac{8}{n-1}$, then both roots are positive, $KT_1 < KT_2$, and the equilibrium is locally asymptotically stable if $KT < KT_1$ or $KT > KT_2$. Figure 4.5 shows the graphs of KT_1 and KT_2 as functions of $\bar{\gamma}$. The stability region is the shaded area.

At $\bar{\gamma} = -1$, for example,

$$KT_{1,2} = \frac{n - 5 \pm \sqrt{n^2 - 10n + 9}}{4}.$$

Proposition 4.18 *In the symmetric case the equilibrium with model* (4.106) *is asymptotically stable if*

$$\begin{aligned} &(a) \text{ there is no delay or } m = 0 \text{ and } (4.111) \text{ holds} \\ &(b)\ m = 1 \text{ with } -\frac{8}{n-1} < \bar{\gamma} \le 1 \\ &\qquad \text{or } \bar{\gamma} = -\frac{8}{n-1} \text{ and } KT \ne 1 \\ &\qquad \text{or } \bar{\gamma} < -\frac{8}{n-1} \text{ and either } KT < KT_1 \text{ or } KT > KT_2, \end{aligned}$$

where KT_{12} *are given in* (4.114).

Notice that from Fig. 4.3, we know that in the concavity region

$$\gamma < -\frac{1}{n-2} \text{ or } \gamma > -\frac{1}{n-1}.$$

The range of $\bar{\gamma}$ is the following:

$$-\infty < \bar{\gamma} \le -\frac{1}{n-3} \text{ and } -\frac{1}{n-1} \le \bar{\gamma} < 1.$$

In addition,

$$\bar{\gamma} = -\frac{8}{n-1} \text{ for } \gamma = -\frac{8}{7n-15}$$

and

$$\bar{\gamma} = 1 \text{ for } \gamma = -\frac{1}{n-1}.$$

From Fig. 4.3, we can conclude that

$$\begin{aligned} -\frac{8}{n-1} \le \bar{\gamma} \le 1 &\text{ if } -1 \le \gamma \le -\frac{8}{7n-15} \text{ if } n > 3 \\ &\text{ or } -\frac{1}{n-1} \le \gamma \le 1, \\ \bar{\gamma} < -\frac{8}{n-1} &\text{ if } -\frac{8}{7n-15} < \gamma < -\frac{1}{n-2} \text{ if } n > 3. \end{aligned}$$

We turn our attention to hyperbolic Cournot oligopolies. With continuously distributed delays, Eq. (4.85) is modified as

$$\dot{q}_k(t) = K_k \left(\sqrt{\frac{\sum_{i \neq k}^{n} q_i^e(t)}{c_k}} - \sum_{i \neq k}^{n} q_i^e(t) - q_k(t) \right) \text{ for } k = 1, 2, \ldots, n. \quad (4.115)$$

where $q_i^e(t)$ is given by Eq. (4.96). The linearized equation becomes

$$\dot{q}_{k\varepsilon}(t) = K_k \left(\sum_{i \neq k}^{n} \gamma_k \int_0^t W(t-s, m_i, T_i) q_{i\varepsilon}(s) ds - q_{k\varepsilon}(t) \right) \text{ for } k = 1, 2, \ldots, n. \quad (4.116)$$

where γ_k is the derivative of the best response function of firm k at the equilibrium. In the discussion on model (4.46), we showed that at the equilibrium,

$$\gamma_k = \frac{C}{2c_k(n-1)} - 1 > -1$$

where $C = \sum_{k=1}^{n} c_k$. Assuming exponential solution forms

$$q_{i\varepsilon}(t) = e^{\lambda t} u_i \text{ for } i = 1, 2, \ldots, n$$

and substituting them into Eq. (4.116), in the same way as in the case of linear oligopolies, we get equations

$$(\lambda + K_k) u_k - K_k \gamma_k \sum_{i \neq k}^{n} \left(1 + \frac{\lambda T_i}{\bar{m}_i} \right)^{-(m_i+1)} u_i = 0 \text{ for } k = 1, 2, \ldots, n.$$

The determinant of which is zero giving the characteristic equation

$$\varphi(\lambda) = \det \begin{pmatrix} \lambda + K_1 & -K_1\gamma_1 \left(1 + \frac{T_2\lambda}{\bar{m}_2}\right)^{-(m_2+1)} & \cdots & -K_1\gamma_1 \left(1 + \frac{T_n\lambda}{\bar{m}_n}\right)^{-(m_n+1)} \\ -K_2\gamma_2 \left(1 + \frac{T_1\lambda}{\bar{m}_1}\right)^{-(m_1+1)} & \lambda + K_2 & \cdots & -K_2\gamma_2 \left(1 + \frac{T_n\lambda}{\bar{m}_n}\right)^{-(m_n+1)} \\ \cdot & \cdot & \cdots & \cdot \\ -K_n\gamma_n \left(1 + \frac{T_1\lambda}{\bar{m}_1}\right)^{-(m_1+1)} & -K_n\gamma_n \left(1 + \frac{T_2\lambda}{\bar{m}_2}\right)^{-(m_2+1)} & \cdots & \lambda + K_n \end{pmatrix} = 0.$$

Similarly to the linear case, introduce the following notation:

$$\boldsymbol{D} = diag \left(\lambda + K_1 \left[1 + \gamma_1 \left(1 + \frac{T_1\lambda}{\bar{m}_1} \right) \right]^{-(m_1+1)}, \ldots, \lambda + K_n \left[1 + \gamma_n \left(1 + \frac{T_n\lambda}{\bar{m}_n} \right) \right]^{-(m_n+1)} \right)$$

$$\boldsymbol{a} = (K_1\gamma_1, \ldots, K_n\gamma_n)^T$$

and

$$\boldsymbol{b}^T = \left(-\left(1 + \frac{T_1\lambda}{\bar{m}_1}\right)^{-(m_1+1)}, \ldots, -\left(1 + \frac{T_n\lambda}{\bar{m}_n}\right)^{-(m_n+1)} \right)$$

to see that

$$\begin{aligned}
\varphi(\lambda) &= \det\left(\boldsymbol{D} + \boldsymbol{a}\boldsymbol{b}^T\right), \\
&= \det(\boldsymbol{D}) \det\left(\boldsymbol{I} + \boldsymbol{D}^{-1}\boldsymbol{a}\boldsymbol{b}^T\right) \\
&= \prod_{k=1}^{n} \left[\lambda + K_k\left(1 + \gamma_k\left(1 + \tfrac{\lambda T_k}{\bar{m}_k}\right)\right)^{-(m_k+1)}\right] \left[1 - \sum_{k=1}^{n} \frac{K_k\gamma_k\left(1+\frac{\lambda T_k}{\bar{m}_k}\right)^{-(m_k+1)}}{\lambda + K_k\left(1+\gamma_k\left(1+\frac{\lambda T_k}{\bar{m}_k}\right)^{-(m_k+1)}\right)}\right] \\
&= 0.
\end{aligned}$$

If the kth term is zero in the first product, then

$$(\lambda + K_k)\left(1 + \frac{\lambda T_k}{\bar{m}_k}\right)^{m_k+1} + K_k\gamma_k = 0. \tag{4.117}$$

This is a polynomial of degree $m_k + 2$, and similarly to Eq. (4.99), it can be proved that the roots are negative real or with negative real parts if $|\gamma_k| < 1$.

Proposition 4.19 *The equilibrium with model* (4.115) *is locally asymptotically stable if the roots of equations* (4.117) *and roots of equation*

$$1 - \sum_{k=1}^{n} \frac{K_k\gamma_k}{(\lambda + K_k)\left(1 + \frac{\lambda T_k}{\bar{m}_k}\right)^{m_k+1} + K_k\gamma_k} = 0 \tag{4.118}$$

are negative real or with negative real parts.

For the sake of mathematical simplicity, consider the special case when $c_1 = \cdots = c_n = c$. Then for all k,

$$\gamma_k = \frac{2-n}{2(n-1)} = \bar{\gamma}'$$

and clearly

$$-\frac{1}{2} < \bar{\gamma}' \leq 0 \text{ for all } n \geq 2.$$

If in addition, we have the symmetric case,

$$T_1 = \cdots = T_n = T,\ m_1 = \cdots = m_n = m \text{ and } K_1 = \cdots = K_n = K,$$

then Eqs. (4.117) and (4.118) become (4.109) and (4.110) with $\bar{\gamma}$ being replaced by $\bar{\gamma}'$. So we can use Proposition 4.18 for this case.

In the cases of no delay and $m = 0$, the stability condition is (4.111),

$$-1 < \bar{\gamma}' < \frac{1}{n-1}$$

which holds for all $n \geq 2$.

In the case of $m = 1$, notice first that

$$\bar{\gamma}' = \frac{2-n}{2(n-1)} > -\frac{8}{n-1} \text{ for } n < 18$$

in which cases, there is no instability region. If $n = 18$, then $\bar{\gamma}' = -8/17$ and $KT_1 = KT_2 = 1$ so equilibrium is locally asymptotically stable if $KT \neq 1$. Assume next that $n \geq 19$, then

$$KT_{12} = \frac{n - 10 \pm \sqrt{n^2 - 20n + 36}}{8} \tag{4.119}$$

and the equilibrium is locally asymptotically stable if $KT < KT_1$ or $KT > KT_2$, when $KT_1 < KT_2$.

Proposition 4.20 *In the symmetric case, the equilibrium with model* (4.115) *is asymptotically stable if*

(a) there is no delays or $m = 0$ with all parameter values
(b) $m = 1$ with $n < 18$
or $n = 18$ with $KT \neq 1$
or $n \geq 19$ and $KT < KT_1$ or $KT > KT_2$,

where $KT_{1,2}$ are given by (4.119).

In the case of hyperbolic Bertrand oligopoly, the dynamic equations are

$$\dot{p}_k(t) = K_k \left[R_k \left(\mathbf{p}^e_{-k}(t) \right) - p_k(t) \right] \tag{4.120}$$

where for $i \neq k$, $p^e_i(t)$ is obtained similarly as $q^e_i(t)$ in (4.96). The best response function of firm k was given in Eq. (4.57) with derivative with respect to p_j $(j \neq k)$,

$$\gamma_{kj} = -\frac{b_{kj}^{-1}}{2p_j^2} \sqrt{-\frac{c_k b_{kk}^{-1}}{\left(\sum_{j \neq k} b_{kj}^{-1} \frac{1}{p_j} \right)^3}} \tag{4.121}$$

as shown in Eq. (4.89). Linearizing the dynamic equation gives the system

$$\dot{p}_{k\varepsilon}(t) = K_k \left[\sum_{i \neq k}^{n} \gamma_{ki} \int_0^t W(t-s, m_i, T_i) p_{i\varepsilon}(s) ds - p_{k\varepsilon}(t) \right]$$

and substituting the exponential solution forms

$$p_{k\varepsilon}(t) = e^{\lambda t} u_k \text{ for } k = 1, 2, \ldots, n$$

and letting $t \to \infty$ in the integrals, we have

$$(\lambda + K_k)\, u_k - K_k \sum_{i \neq k}^{n} \gamma_{ki} \left(1 + \frac{\lambda T_i}{\bar{m}_i}\right)^{-(m_i+1)} u_i = 0 \text{ for } k = 1, 2, \ldots, n.$$

The determinant of this system is zero giving the characteristic polynomial

$$\varphi(\lambda) = \det \begin{pmatrix} \lambda + K_1 & -K_1\gamma_{12}\left(1 + \frac{T_2\lambda}{\bar{m}_2}\right)^{-(m_2+1)} & \cdots & -K_1\gamma_{1n}\left(1 + \frac{T_n\lambda}{\bar{m}_n}\right)^{-(m_n+1)} \\ -K_2\gamma_{21}\left(1 + \frac{T_1\lambda}{\bar{m}_1}\right)^{-(m_1+1)} & \lambda + K_2 & \cdots & -K_2\gamma_{2n}\left(1 + \frac{T_n\lambda}{\bar{m}_n}\right)^{-(m_n+1)} \\ \cdot & \cdot & \cdots & \cdot \\ -K_n\gamma_{n1}\left(1 + \frac{T_1\lambda}{\bar{m}_1}\right)^{-(m_1+1)} & -K_n\gamma_{n2}\left(1 + \frac{T_2\lambda}{\bar{m}_2}\right)^{-(m_2+1)} & \cdots & \lambda + K_n \end{pmatrix}$$

which is similar to the fixed delay case (4.93).

Proposition 4.21 *The equilibrium with model* (4.120) *is locally asymptotically stable if all roots of $\varphi(\lambda)$ are negative or complex with negative real parts.*

In general, no simple stability conditions can be given; therefore, we consider the symmetric case of

$$K_1 = \cdots = K_n = K, \; c_1 = \cdots = c_n = c \text{ and } \Theta_{kj} = \Theta \text{ for all } k \text{ and } j.$$

From the previous section, we know that in this case,

$$\gamma_{kj} = \bar{\gamma} = \frac{1}{2(n-1)}.$$

If, in addition,

$$T_1 = \cdots = T_n = T \text{ and } m_1 = \cdots = m_n = m,$$

then this eigenvalue equation is identical to (4.98), when $-\gamma/2$ is replaced by $\bar{\gamma}$. So Eqs. (4.99) and (4.100) still hold which become in the symmetric case Eqs. (4.109) and (4.110). Notice first that (4.111) holds with this $\bar{\gamma}$ value, so without delay and with $m = 0$ the equilibrium is locally asymptotically stable. If $m = 1$, then, since $\bar{\gamma} > 0$, (4.112) is the stronger condition, in which case it was shown that the equilibrium is locally asymptotically stable for all $KT > 0$.

Proposition 4.22 *In the symmetric case, the equilibrium with model* (4.120) *is locally asymptotically stable without delay and for $m = 0$ and $m = 1$.*

The cases of mixed oligopolies can be examined similarly to the case shown at the end of Sect. 4.1 where the derivatives γ_1, γ_2, γ_3 and γ_4 are different. The block matrices of the Jacobians without and with delays are similar to that case including the additional factors

$$\left(1+\frac{\lambda T_k}{\bar{m}_k}\right)^{-(m_k+1)}$$

in all off-diagonal terms which make the analysis even more complicated than the cases of models (4.64) and (4.68). Therefore, the details are not discussed here.

4.4 Supplementary Notes and Discussions

The theory of oligopoly was originated from the pioneering work of Cournot (1838) and since then a huge literature was developed. Among the large number of books and chapters devoted to this field, we mention Burger (1959), Okuguchi (1976), Friedman (1977), Friedman (1981), Okuguchi and Szidarovszky (1999) and Bischi et al. (2010). Most books on game theory devote at least a special section to oligopoly such as Szep and Forgo (1985), Matsumoto and Szidarovszky (2015e) among many others.

The introduction of this chapter is based on Matsumoto and Szidarovszky (2011a, b) with a new material about mixed hyperbolic oligopolies. Section 4.1 uses some selected parts of Matsumoto and Szidarovszky (2011a, 2014e, 2011c) with some added new material about the delayed dynamics. Sections 4.2 and 4.3 contain mainly newly developed material.

The classical paper of Theocharis (1959) is the first important result on the stability of discrete time oligopolies. Among the early works on stability, we mention Hahn (1962), Al-Nowaihi and Levine (1985), Furth (1986), Hadar (1966), Fisher (1961). Puu (2003) gives an excellent summary of nonlinear phenomena, and Puu and Sushko (2002) edited a collection of papers on dynamic oligopoly theory. Chiarella et al. (2013) discusses optimal strategies in hyperbolic oligopolies when the price function has a random error and each firm uses certainty equivalent for optimality. Continuously distributed delays in dynamic economics were introduced by Invernizzi and Medio (1991) and were used for dynamic oligopolies by Chiarella and Khomin (1996). If no analytic method is available, then either computer simulations or laboratory experiment can be used to examine the asymptotical behavior of the state trajectories and determine the typical types of attractors and bifurcation diagrams (Kopel 1996; Kubicek and Marek 1986; Cox and Walker 1998). Cournot and Bertrand oligopolies were compared by Cheng (1985), Okuguchi (1984) and Häckner (2000) among others.

Chapter 5
Learning in Monopolies and Oligopolies

In the previous chapters, we always assumed that the firms had complete knowledge about the types and the parameters of all functions involved in the mathematical models. In reality, however, this is not always the case. The firms are able to assess their actual costs by accurate bookkeeping, and then, the function forms of price function can be obtained by certain function approximation method, like least squares based on actual market data. The case of the price function is even more complicated. While the firms know their own output levels for the cost function, in most cases, they can only guess or estimate the industry output which is the independent variable of the price function.

In this chapter, special learning processes will be introduced and examined in which the firms can use current and past market observations to improve their assessments about the unknown parameters of the price function. The continuous updating of these "beliefs" leads to special dynamic processes with both discrete and continuous time scales. Conditions can be derived to guarantee that the sequence of the repeated estimates converges to the true value showing successful learning. If the market data are obtained with some delay and, in addition, their implementation needs time, then the dynamic models have to include time delays as well.

This chapter consists of two sections. First, learning models in monopoly will be discussed, and in the second section, n-firm oligopolies will be investigated. Duopolies are special cases with the choice of $n = 2$.

5.1 Learning in Monopolies

A simple linear monopoly is considered, where q is the output of the firm, the price function is linear, $p(q) = a - bq$ $(a, b > 0)$, and there is a constant marginal cost, c. It is assumed that $a > c$ as in Chap. 2. The profit of the monopoly is given as

A. Matsumoto and F. Szidarovszky, *Dynamic Oligopolies with Time Delays*,
https://doi.org/10.1007/978-981-13-1786-6_5

$$\pi(q) = (a - bq)\,q - cq. \tag{5.1}$$

It is assumed that the firm does not know the true value of a; it has only an estimate of it, which is denoted by a^e. So the believed profit of the firm is

$$\pi^e(q) = \left(a^e - bq\right)q - cq$$

implying that the optimal output level is believed to be

$$q^e = \frac{a^e - c}{2b}, \tag{5.2}$$

so the firm produces this amount and expects the price to be

$$p^e(q^e) = a^e - bq^e = \frac{a^e + c}{2}.$$

However, the actual market price becomes

$$p(q^e) = a - bq^e = \frac{2a - a^e + c}{2}$$

based in relation (5.2). At each time period t, the firm computes the expected price, receives the market price and tries to adjust its estimate a^e of the maximum price based on the discrepancy between them. Notice that

$$p(q^e) - p^e(q^e) = a - a^e, \tag{5.3}$$

so the firm can think in the following way. If $p(q^e) > p^e(q^e)$, then the estimate a^e is too low, so the firm wants to increase it. If $p(q^e) < p^e(q^e)$, then the estimate a^e is too large, so the firm wants to decrease it, and if they are equal, then the firm believes that its estimate a^e is correct. So the change in the estimate a^e is done in proportion to the price difference (5.3), which is equivalent to be done in proportion to $a - a^e$.

In considering discrete time scales, this concept is described by the dynamic equation

$$a^e(t+1) = a^e(t) + ka^e(t)\left[a - a^e(t)\right] \tag{5.4}$$

which is a nonlinear difference equation. Clearly, $a^e = a$ is the only nonzero steady state. The asymptotical behavior of the state trajectory of this system is examined by linearization. Since

$$\frac{\partial a^e(t+1)}{\partial a^e(t)} = 1 - ka$$

at the steady state, the linearized equation becomes

$$a^e_\varepsilon(t+1) = (1 - ka)\,a^e_\varepsilon(t)$$

where $a^e_\varepsilon(t)$ is the difference between $a^e(t)$ and the steady state a. This linear system is asymptotically stable if and only if

$$|1 - ka| < 1$$

that is,

$$ka < 2 = ka_0^*.$$

In this case, the learning process is successful and converges to the true value.

Proposition 5.1 *The steady state with model* (5.4) *is locally asymptotically stable if the speed of adjustment k is less than* $\frac{2}{a}$.

Assume next that there is a delay τ in obtaining the market price information, which is

$$p(q^e(t-\tau)) = \frac{2a - a^e(t-\tau) + c}{2},$$

however the firm does not want to compare this past price data with its own most recent expectation, so it makes the comparison with its expectation at the same time period $t - \tau$. So the dynamic equation (5.4) is modified as the following delay equation

$$a^e(t+1) = a^e(t) + ka^e(t)\left[a - a^e(t-\tau)\right]. \tag{5.5}$$

The dynamic behavior of this system clearly depends on the length τ of the delay. Without delay ($\tau = 0$), the dynamic equation (5.4) is equivalent to the well-known logistic equation by introducing the new variable,

$$z(t) = \frac{ka^e(t)}{1 + ak},$$

when it becomes

$$z(t+1) = (1 + ak)z(t)\left[1 - z(t)\right].$$

This logistic equation can generate wide variety of dynamics ranging from periodic cycle to chaos depending on the value of the factor $1 + ak$ in the cases of locally unstable steady states.

By simple differentiation, we have

$$\frac{\partial a^e(t+1)}{\partial a^e(t)} = 1 + k\left[a - a^e(t-\tau)\right] = 1$$

and

$$\frac{\partial a^e(t+1)}{\partial a^e(t-\tau)} = -ka^e(t) = -ka$$

at the equilibrium. So the linearized version of Eq. (5.5) can be written as

$$a_\varepsilon^e(t+1) = a_\varepsilon^e(t) - kaa_\varepsilon^e(t-\tau). \tag{5.6}$$

Assume first that $\tau = 1$. Then, this equation can be rewritten as a two-dimensional first-order equation without delay:

$$\begin{aligned} x(t+1) &= x(t) - kay(t) \\ y(t+1) &= x(t) \end{aligned} \tag{5.7}$$

where $x(t) = a_\varepsilon^e(t)$ and $y(t) = x(t-1)$. The Jacobian matrix is

$$\boldsymbol{J} = \begin{pmatrix} 1 & -ka \\ 1 & 0 \end{pmatrix}$$

and therefore the characteristic equation becomes

$$\begin{aligned} \varphi(\lambda) &= \det\begin{pmatrix} 1-\lambda & -ka \\ 1 & -\lambda \end{pmatrix} \\ &= \lambda^2 - \lambda + ka = 0. \end{aligned}$$

In Appendix C, sufficient and necessary conditions are given in (C9) to guarantee that the roots of this quadratic equation are inside the unit circle:

$$\begin{aligned} ka &< 1, \\ 1 + ka + 1 &> 0, \\ -1 + ka + 1 &> 0 \end{aligned}$$

which hold if and only if

$$ka < 1 = ka_1^*.$$

That is, this condition is stronger than in the no-delay case.

Assume next $\tau = 2$. Then, Eq. (5.6) can be rewritten as a three-dimensional system of first-order equations without delay:

$$\begin{aligned} x(t+1) &= x(t) - kaz(t), \\ y(t+1) &= x(t), \\ z(t+1) &= y(t), \end{aligned} \tag{5.8}$$

where $y(t) = x(t-1)$ and $z(t) = x(t-2)$. The Jacobian matrix is

$$J = \begin{pmatrix} 1 & 0 & -ka \\ 1 & 0 & 0 \\ 0 & 1 & 0 \end{pmatrix}$$

and the corresponding characteristic equation is

$$\det(\boldsymbol{J} - \lambda \boldsymbol{I}) = \lambda^3 - \lambda^2 + ka = 0.$$

This is a cubic equation with coefficients

$$b_1 = -1, \ b_2 = 0 \text{ and } b_3 = ak$$

in the notation of Eq. (C8) of Appendix C. The sufficient and necessary conditions that all roots are inside the unit circle are as follows (see (C10) in Appendix$C$):

$$\begin{gathered} 1 + b_1 + b_2 + b_3 > 0, \\ 1 - b_1 + b_2 - b_3 > 0, \\ 1 - b_2 + b_1 b_3 - b_3^2 > 0, \\ 3 - b_2 > 0, \end{gathered}$$

which are in our case,

$$ak > 0, \ ak < 2, \ ak + (ak)^2 < 1 \text{ and } 0 < 3.$$

The first and last conditions hold; the third is satisfied if $ka < (\sqrt{5} - 1)/2 \simeq 0.618$ which is stronger than the second condition. Therefore, the steady state is locally asymptotically stable with $\tau = 2$ if

$$ka < \frac{\sqrt{5} - 1}{2} = ka_2^*.$$

Consider next the case of $\tau = 3$. Then similarly to the previous case, it is easy to see that the characteristic equation is

$$\lambda^4 - \lambda^3 + ak = 0.$$

In this case, $b_1 = -1$, $b_2 = b_3 = 0$ and $b_4 = ak$. The stability conditions now are much more complicated, as shown in (C11) of Appendix C:

$$\begin{gathered} 1 - b_4 > 0, \\ 3 + 3b_4 > b_2, \\ 1 + b_1 + b_2 + b_3 + b_4 > 0, \\ 1 - b_1 + b_2 - b_3 + b_4 > 0, \\ (1 - b_4)\left(1 - b_4^2\right) - b_2 (1 - b_4)^2 + (b_1 - b_3)(b_3 - b_1 b_4) > 0. \end{gathered}$$

If $ak < 1$, then the first four conditions are satisfied, and the last condition is a cubic inequality in variable ka:

$$f(ka) = (ka)^3 - (ka)^2 - 2(ka) + 1 > 0.$$

Notice that

$$f(-1) = 1,\ f(0) = 1,\ f(1) = -1,\ f(-\infty) = -\infty \text{ and } f(\infty) = \infty$$

implying that there are three real roots. They are in intervals $(-\infty, -1)$, $(0, 1)$ and $(1, \infty)$. Since from the first four conditions, $ka < 1$ should hold, we select the root, $ka \simeq 0.445$ and conclude that the stability condition is

$$ka < 0.445 = ka_3^*$$

In the general case, we can derive the characteristic equation similarly to the case of continuous dynamics; when instead of the exponential form of the solution, we are looking for the solution in the form of

$$a_\varepsilon^e(t) = \lambda^t u.$$

Substituting it into Eq. (5.6), we have

$$\lambda^{t+1} u = \lambda^t u - ka\lambda^{t-\tau} u$$

and by dividing both sides by $\lambda^{t-\tau} u$ we get the simplified equation

$$\lambda^{\tau+1} - \lambda^\tau + ka = 0 \tag{5.9}$$

which is the straightforward generalization of the special equations with $\tau = 1, 2$ and 3.

Proposition 5.2 *The steady state with model* (5.6) *is locally asymptotically stable if* $ka < ka_\tau^*$, *where*

$$ka_0^* = 2,\ ka_1^* = 1,\ ka_2^* = \frac{\sqrt{5}-1}{2} \simeq 0.618 \text{ and } ka_3^* \simeq 0.445042.$$

At these critical values, Neimark–Sacker bifurcation occurs.

If $\tau > 3$, then the critical values of ka can be obtained by the Samuelson or the Cohn–Schur stability conditions (Gandolfo 2009) which are much more complicated than in the cases shown above. In all cases, the asymptotical stability of the steady state shows that the learning process is successful.

If continuous time scales are assumed, then in Eq. (5.5) the difference $a^e(t+1) - a^e(t)$ is replaced by the derivative $\dot{a}^e(t)$ to get the continuous time equation,

$$\dot{a}^e(t) = ka^e(t)\left[a - a^e(t-\tau)\right], \tag{5.10}$$

where τ is a fixed delay. The linearized version of this equation is the following:

$$\dot{a}^e_\varepsilon(t) = -kaa^e_\varepsilon(t-\tau), \tag{5.11}$$

since

$$\frac{\partial \dot{a}^e(t)}{\partial a^e(t)} = 0 \text{ and } \frac{\partial \dot{a}^e(t)}{\partial a^e(t-\tau)} = -ka$$

at the steady state. The characteristic polynomial can be obtained by assuming exponential solution form

$$a^e_\varepsilon(t) = e^{\lambda t}u$$

and substituting it into Eq. (5.11),

$$\lambda e^{\lambda t}u + kae^{\lambda(t-\tau)}u = 0$$

and after simplifying by $e^{\lambda t}u$, we have

$$\lambda + kae^{-\lambda\tau} = 0. \tag{5.12}$$

If $\tau = 0$, when no delay is present, then this equation becomes $\lambda + ka = 0$ with negative eigenvalue, $\lambda = -ka < 0$.

Proposition 5.3 *The continuous model without delay is always locally asymptotically stable, so learning is successful.*

As the length of the delay increases from zero, at a certain value of τ stability might be lost, when at least one eigenvalue has zero real part: $\lambda = iv$ $(v > 0)$. Substituting this solution into Eq. (5.12), we have

$$iv + ka(\cos v\tau - i\sin v\tau) = 0$$

and by separating the real and imaginary parts,

$$ka\cos v\tau = 0$$

and

$$v - ka\sin v\tau = 0.$$

Therefore, $\cos v\tau = 0$, $\sin v\tau = 1$ and $v = ka$ implying that

$$v\tau = \frac{\pi}{2} + 2n\pi,$$

so there are infinitely many solutions,

$$\tau_n^* = \frac{1}{ka}\left(\frac{\pi}{2} + 2n\pi\right) \quad (n = 0, 1, 2, \ldots). \tag{5.13}$$

The direction of the stability switches at the critical values can be obtained by idea of Hopf bifurcation. Select τ as the bifurcation parameter, and consider the eigenvalue as the function of τ, $\lambda = \lambda(\tau)$. By implicitly differentiating equation (5.12) with respect to τ, we get a simple equation for λ' as follows:

$$\lambda'(\tau) + kae^{-\lambda\tau}\left(-\lambda'(\tau)\tau - \lambda(\tau)\right) = 0,$$

so

$$\lambda'(\tau) = \frac{kae^{-\lambda\tau}\lambda(\tau)}{1 - kae^{-\lambda\tau}\tau}.$$

However from Eq. (5.12), we know that $kae^{-\lambda\tau} = -\lambda$, so

$$\lambda'(\tau) = \frac{-\lambda^2(\tau)}{1 + \lambda(\tau)\tau}.$$

At $\lambda = iv$,

$$\lambda'(\tau) = \frac{v^2}{1 + iv\tau}\frac{1 - iv\tau}{1 - iv\tau} = \frac{v^2 - iv^3\tau}{1 + v^2\tau^2}$$

with positive real part

$$\mathrm{Re}\left[\lambda'(\tau)\right] = \frac{v^2}{1 + v^2\tau^2}$$

showing that at the critical values of τ at least one pair of eigenvalues changes the sign of its real part from negative to positive. Therefore at the smallest critical value τ_0^*, stability is lost and it cannot be regained later.

Proposition 5.4 *The steady state with model* (5.10) *is locally asymptotically stable if $\tau < \tau_0^*$, locally unstable as $\tau > \tau_0^*$, and Hopf bifurcation occurs at $\tau = \tau_0^*$, and a stable limit cycle exist for $\tau > \tau_0^*$ where*

$$\tau_0^* = \frac{\pi}{2ka}.$$

In the case when the steady state is asymptotically stable, the learning process is successful. We selected $ka = 2$ and illustrated the bifurcation diagram with respect to τ in Fig. 5.1a, where it is clear that the steady state loses stability at τ_0^* and bifurcates to a cyclic oscillation as $\tau > \tau_0^*$. The vertical line at $\tau_a > \tau_0^*$ crosses the stability

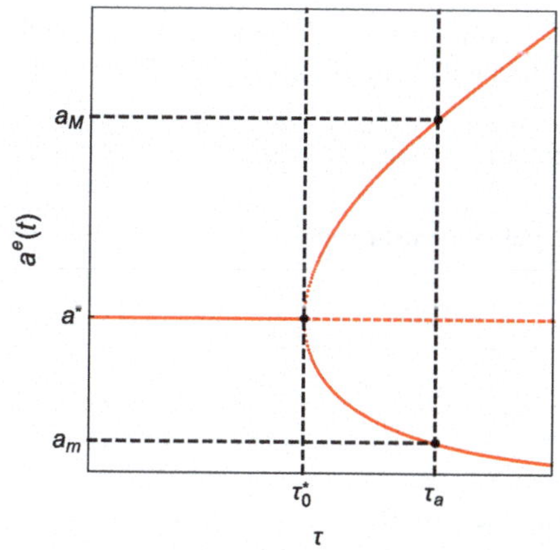

(a) Bifurcation diagram

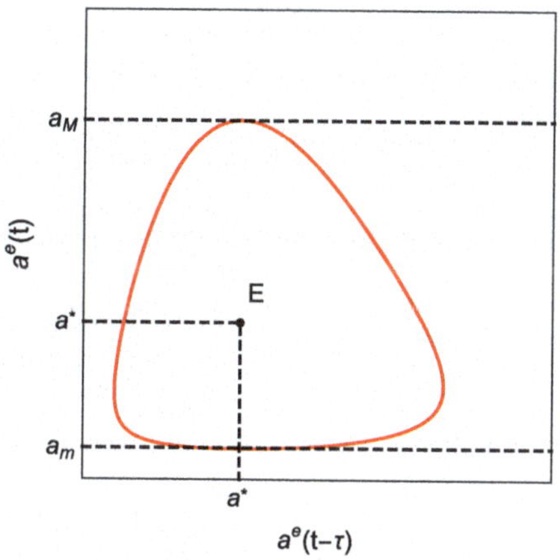

(b) Limit cycle with $\tau = \tau_a$

Fig. 5.1 Cyclic oscillations for $\tau > \tau^*$

switching curve at two points with ordinates a_m and a_M, which are the minimum and maximum values of the trajectory as it is also shown in Fig. 5.1b by the limit cycle at $\tau = \tau_a$.

Consider next another delay adjustment process which is a combination of Eq. (5.10) with the no-delay model:

$$\dot{a}^e(t) = ka^e(t) \left\{ a - \left[\omega a^e(t) + (1-\omega) a^e(t-\tau) \right] \right\} \tag{5.14}$$

where $0 \le \omega \le 1$. If $\omega = 1$, then the no-delay equation is obtained, and as $\omega = 0$, it is simplified to Eq. (5.10). Notice that

$$\frac{\partial \dot{a}^e(t)}{\partial a^e(t)} = -ka\omega \text{ and } \frac{\partial \dot{a}^e(t)}{\partial a^e(t-\tau)} = -ka(1-\omega)$$

at the steady state, so the linearized equation becomes

$$\dot{a}^e_\varepsilon(t) + ka\omega a^e_\varepsilon(t) + ka(1-\omega) a^e_\varepsilon(t-\tau) = 0.$$

By substituting an exponential solution form

$$a^e_\varepsilon(t) = e^{\lambda t} u,$$

we can see that the characteristic equation is the following:

$$\lambda + ka\omega + ka(1-\omega) e^{-\lambda\tau} = 0. \tag{5.15}$$

Without delay, the steady state is locally asymptotically stable, and stability switch might occur when $\lambda = iv$ ($v > 0$) which is then substituted into Eq. (5.15) to have

$$iv + ka\omega + ka(1-\omega)(\cos v\tau - i \sin v\tau) = 0.$$

Separating the real and imaginary parts, we get two equations

$$\begin{aligned} ka(1-\omega)\cos v\tau &= -ka\omega, \\ ka(1-\omega)\sin v\tau &= v. \end{aligned} \tag{5.16}$$

Adding the squares of these equations yields

$$k^2a^2(1-\omega)^2 = k^2a^2\omega^2 + v^2,$$

so

$$v^2 = k^2a^2(1-2\omega).$$

If $\omega \geq 1/2$, then no solution exists implying that there is no stability switch, and the steady state is locally asymptotically stable. Otherwise,

$$v = ka\sqrt{1-2\omega} > 0$$

and therefore from (5.16) we have

$$\tau_n^{**} = \frac{1}{ka\sqrt{1-2\omega}}\left[-\sin^{-1}\left(\frac{\sqrt{1-2\omega}}{1-\omega}\right) + (2n+1)\pi\right] \quad (n = 0, 1, 2, \ldots). \tag{5.17}$$

The directions of stability switches are determined similarly to the previous case. Assume $\lambda = \lambda(\tau)$; then, implicit differentiation of equation (5.15) shows that

$$\lambda'(\tau) + ka(1-\omega)e^{-\lambda(\tau)\tau}\left[-\lambda'(\tau)\tau - \lambda(\tau)\right] = 0$$

so

$$\lambda'(\tau) = \frac{ka(1-\omega)e^{-\lambda(\tau)\tau}\lambda(\tau)}{1 - ka(1-\omega)e^{-\lambda(\tau)\tau}\tau}.$$

However from the characteristic equation

$$ka(1-\omega)e^{-\lambda(\tau)\tau} = -\lambda(\tau) - ka\omega$$

implying that

$$\lambda'(\tau) = \frac{-\lambda^2 - ka\omega\lambda(\tau)}{1 + \lambda(\tau)\tau + ka\omega\tau}.$$

At $\lambda = iv$, it becomes

$$\lambda'(\tau) = \frac{v^2 - ka\omega iv}{1 + ka\omega\tau + iv\tau} \frac{1 + ka\omega\tau - iv\tau}{1 + ka\omega\tau - iv\tau}$$

the real part of which is positive,

$$\mathrm{Re}\left[\lambda'(\tau)\right] = \frac{v^2}{(1 + ka\omega\tau)^2 + (v\tau)^2} > 0.$$

So we derived the following result.

Proposition 5.5 *If* $\omega \geq \frac{1}{2}$*, then the steady state of process* (5.14) *is locally asymptotically stable for all* $\tau > 0$*. If* $\omega < \frac{1}{2}$*, then it is asymptotically stable if* $\tau < \tau_0^{**}$*, loses stability at* $\tau = \tau_0^{**}$ *and generates a limit cycle via Hopf bifurcation if* $\tau > \tau_0^{**}$ *where*

$$\tau_0^{**} = \frac{1}{ka\sqrt{1-2\omega}}\left[\pi - \sin^{-1}\left(\frac{\sqrt{1-2\omega}}{1-\omega}\right)\right].$$

Therefore, the delayed learning process (5.14) is successful if either $\omega \geq 1/2$ or if $\omega < 1/2$ and the delay is less than τ_0^{**}.

Consider next the two-delay extension of the previously examined models:

$$\dot{a}^e(t) = ka^e(t)\left[a - \omega a^e(t - \tau_1) - (1 - \omega)\, a^e(t - \tau_2)\right] \tag{5.18}$$

where $\tau_1, \tau_2 > 0$ are the delays and $0 \leq \omega \leq 1$. The linearized equation has the form

$$\dot{a}^e_\varepsilon(t) + ka\left[\omega a^e_\varepsilon(t - \tau_1) + (1 - \omega)\, a^e_\varepsilon(t - \tau_2)\right] = 0.$$

Substitution of the exponential solution form

$$a^e_\varepsilon(t) = e^{\lambda t} u$$

provides the characteristic equation

$$\lambda + ka\omega e^{-\lambda\tau_1} + ka(1 - \omega)e^{-\lambda\tau_2} = 0. \tag{5.19}$$

This has the same form as (A6) of Appendix A with

$$P_0(\lambda) = \lambda,\ \ P_1(\lambda) = ka\omega \text{ and } P_2(\lambda) = ka(1 - \omega).$$

In order to apply the method derived in Appendix A for two-delay equations, we need to check four conditions:

(i) $\deg\left[P_0(\lambda)\right] = 1 > \max\left\{\deg\left[P_1(\lambda)\right],\ \deg\left[P_2(\lambda)\right]\right\} = 0.$
(ii) $P_0(0) + P_1(0) + P_2(0) = ka \neq 0.$

(iii) Polynomials $P_0(\lambda)$, $P_1(\lambda)$ and $P_2(\lambda)$ have no common roots.
(iv)

$$\lim_{\lambda\to\infty}\left[\left|\frac{P_1(\lambda)}{P_0(\lambda)}\right|+\left|\frac{P_2(\lambda)}{P_0(\lambda)}\right|\right]=0<1.$$

Notice first that Eq. (5.19) can be rewritten as

$$1+a_1(\lambda)e^{-\lambda\tau_1}+a_2(\lambda)e^{-\lambda\tau_2}=0 \tag{5.20}$$

with

$$a_1(\lambda)=\frac{ka\omega}{\lambda} \text{ and } a_2(\lambda)=\frac{ka(1-\omega)}{\lambda}.$$

Clearly with $\lambda=iv$ $(v>0)$,

$$|a_1(iv)|=\frac{ka\omega}{v} \text{ and } |a_2(iv)|=\frac{ka(1-\omega)}{v},$$

therefore the domain of feasible values of v are given by relations (A.11) and (A.12), which are in our case as follows:

$$\frac{ka\omega}{v}+\frac{ka(1-\omega)}{v}\geq 1,$$
$$-1\leq\frac{ka\omega}{v}-\frac{ka(1-\omega)}{v}\leq 1.$$

From the first condition, we have $v\leq ka$, and from the second, we get $-v\leq ka(2\omega-1)\leq v$. We can assume that $\tau_1<\tau_2$ because of the symmetry in τ_1 and τ_2. If the firm gives higher weight to more recent delay, then $\omega\geq 1/2$. Therefore, the feasible domain of v is the following,

$$ka(2\omega-1)\leq v\leq ka.$$

From (A.13) and (A.14), we have

$$\Theta_1(v)=\cos^{-1}\left(\frac{1+|a_1(iv)|^2-|a_2(iv)|^2}{2\,|a_1(iv)|}\right)=\cos^{-1}\left(\frac{v^2+(ka)^2\,(2\omega-1)}{2kav\omega}\right)$$

and

$$\Theta_2(v)=\cos^{-1}\left(\frac{1+|a_2(iv)|^2-|a_1(iv)|^2}{2\,|a_2(iv)|}\right)=\cos^{-1}\left(\frac{v^2+(ka)^2\,(1-2\omega)}{2kav(1-\omega)}\right).$$

The stability switching curves are therefore given from Eqs. (A.15) and (A.16) as

$$
\begin{aligned}
\tau_1^{n\pm}(v) &= \frac{1}{v}\left[\frac{3\pi}{2} + (2n-1)\pi \pm \Theta_1(v)\right] \\
\tau_2^{m\pm}(v) &= \frac{1}{v}\left[\frac{3\pi}{2} + (2m-1)\pi \mp \Theta_1(v)\right],
\end{aligned}
\tag{5.21}
$$

since

$$
\arg\left(\frac{ka\omega}{iv}\right) = \arg\left(-i\frac{ka\omega}{v}\right) \text{ and } \arg\left(\frac{ku(1-\omega)}{iv}\right) = \arg\left(-i\frac{ka(1-\omega)}{v}\right).
$$

So we have the complete description of the stability switching curves from (A.17) and (A.18).

Proposition 5.6 *The stability switching curves are the union of the curves* $\left(\tau_1^{n\pm}(v), \tau_2^{m\pm}(v)\right)$, $v \in [ka(2\omega - 1), ka]$ *where with fixed values of n, the segments*

$$
C^+(n,m) = \{\tau_1^{n+}(v), \tau_2^{m+}(v)\}
$$

and

$$
C^-(n,m) = \{\tau_1^{n-}(v), \tau_2^{m-}(v)\}
$$

form continuous curves as m increases.

Figure 5.2 shows the partition curves with $n = 0$ and $m = 0$, 1 and 2 with parameter selection $\omega = 0.8$ and $a = k = 1$. The aspect ratio is adjusted to $1 \leq \tau_1 \leq 2.8$ and $0 \leq \tau_2 \leq 19$. The positive sloping line is the $\tau_2 = \tau_1$ line, under which the assumption $\tau_1 < \tau_2$ is violated, so that region is eliminated from further considerations. The red curves are $C^+(n,m)$, and the blue curves are $C^-(n,m)$. The green dots are the initial points, and the black dots are the end points of the segments. The two curves are connected at these points. From (5.21), we see that the red and blue segments shift upward when m increases. The τ_1^m value gives the minimum τ_1 values of the segments $C^-(n,m)$, and τ_1^M gives the maximum τ_1 values in the $C^+(n,m)$ segments. In order to keep τ_2 positive, $C^+(0,0)$ is shown only for $\tau_1 < \tau_1^0 \simeq 2.35$ in our case, when $\tau_1^m \simeq 1.493$ and $\tau_1^M \simeq 2.733$. These values are numerically determined by solving the first-order conditions

$$
\frac{\partial \tau^-}{\partial v} = 0 \text{ and } \frac{\partial \tau^+}{\partial v} = 0
$$

for v, and the corresponding v values become $v_m \simeq 0.956$ and $v_M \simeq 0.621$.

We therefore observe that the steady state is locally asymptotically stable if $\tau_1 = 0$ and $\tau_2 \geq 0$ as well as if $0 < \tau_1 < \tau_1^m$ and $\tau_2 \geq \tau_1$. The transient dynamics however depends on the value of the delays. Figure 5.3 shows three cases with $\tau_1 = 1.1$ and $\tau_2^R = 1.1$, $\tau_2^B = 2.2$ and $\tau_2^G = 3.3$, where R, B and G stand for red, blue and green colors. It is well demonstrated how the amplitudes of the oscillating curves vary as well as how the relative order of the amplitudes of the three cases is different.

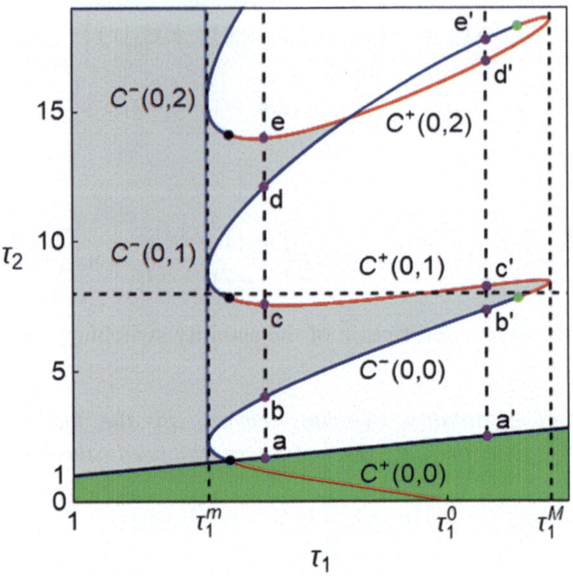

Fig. 5.2 Partition curve in the (τ_1, τ_2) plane

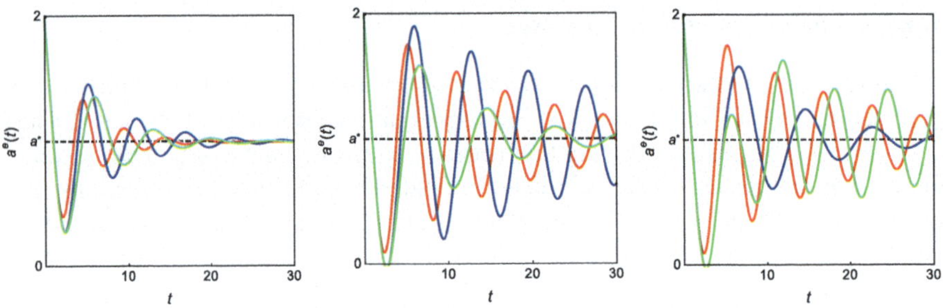

(a) $R > G > B$ (b) $G > R > B$ (c) $B > R > G$

Fig. 5.3 Harmless delays affect convergence speed

The direction of the stability switches can be determined based on Theorem A.1 of Appendix A. Keeping its notation, we can see that

$$R_2 I_1 = \text{Re}\left[a_2(iv)e^{-iv\tau_2}\right]\text{Im}\left[a_1(iv)e^{-iv\tau_1}\right]$$

and

$$R_1 I_2 = \text{Re}\left[a_1(iv)e^{-iv\tau_1}\right]\text{Im}\left[a_2(iv)e^{-iv\tau_2}\right],$$

and therefore, it is easy to verify that

$$\Phi = R_2 I_1 - R_1 I_2 = \text{Im}\left[a_1(iv)a_2(-iv)e^{iv(\tau_2-\tau_1)}\right].$$

In our case

$$\begin{aligned}\Phi &= \mathrm{Im}\left[\frac{ka\omega}{iv}\frac{ka(1-\omega)}{-iv}\left(\cos v\,(\tau_2-\tau_1)+i\sin v\,(\tau_2-\tau_1)\right)\right]\\ &= \frac{(ka)^2\omega(1-\omega)}{v^2}\sin v\,(\tau_2-\tau_1)\,,\end{aligned}$$

so the sign of Φ is the same as that of $\sin v\,(\tau_2-\tau_1)$. Notice first that on segments $C^+(n,m)$

$$\sin v\,(\tau_2-\tau_1) = \sin\left(-\cos^{-1}A-\cos^{-1}B\right)$$

with

$$A=\frac{v^2+(ka)^2(2\omega-1)}{2kav\omega}>0 \text{ and } B=\frac{v^2-(ka)^2(2\omega-1)}{2kav(1-\omega)}.$$

If $v\geq ka\sqrt{2\omega-1}$, then $B\geq 0$, and if $v<ka\sqrt{2\omega-1}$, then $B<0$. In addition, both $\cos^{-1}A$ and $\cos^{-1}B$ are between 0 and π and

$$\begin{aligned}\sin v\,(\tau_2-\tau_1) &= -\sin(\cos^{-1}A)\cos(\cos^{-1}B)-\cos(\cos^{-1}A)\sin(\cos^{-1}B)\\ &= -B\sqrt{1-A^2}-A\sqrt{1-B^2},\end{aligned}$$

which is positive if $B<0$ and $-B\geq A$. It means that

$$\frac{-v^2+(ka)^2(2\omega-1)}{2kav(1-\omega)}\geq\frac{v^2+(ka)^2(2\omega-1)}{2kav\omega}$$

being equivalent to relation $v\leq ka(2\omega-1)$ which is impossible unless v is at the end point of the segment. So $\Phi<0$ on $C^+(n,m)$.

Consider next the segments $C^-(n,m)$, where

$$\begin{aligned}\sin v\,(\tau_2-\tau_1) &= \sin\left(\cos^{-1}A+\cos^{-1}B\right)\\ &= \sin(\cos^{-1}A)\cos(\cos^{-1}B)+\cos(\cos^{-1}A)\sin(\cos^{-1}B)\\ &= B\sqrt{1-A^2}+A\sqrt{1-B^2},\end{aligned}$$

being positive if $B\geq 0$ or if $B<0$ and $A>-B$ or $v>ka\sqrt{2\omega-1}$, which is true. Thus,

$$\Phi<0 \text{ on } C^+(n,m) \text{ and } \Phi>0 \text{ on } C^-(n,m).$$

From Theorem A.1 of Appendix A, we can conclude the following facts.

Proposition 5.7 *As point (τ_1,τ_2) moves from the right to the left on crossing segment $C^+(n,m)$, a pair of eigenvalues changes the sign of its real part from positive to negative, and on segment $C^-(n,m)$, a pair of eigenvalues changes the sign of its real part from negative to positive.*

In Fig. 5.2 by moving up on the two vertical broken lines stability is lost at points $c,\ e,\ c'$ and stability is regained at points, $b,\ d$ and b'. In order to understand what happens at points d' and e', we assert that all eigenvalues are simple. The lengthy proof shows that any multiple root is at a stationary point of both τ_1^+ and τ_2^+ which is impossible. Consider now point b' in Fig. 5.2, where stability is regained, so all eigenvalues have negative real parts. Moving up to point c', one pair of eigenvalues will have positive real part, and at point d', another pair of eigenvalues changes its real part from negative to positive, and then at point e', only one of them changes back its real part to negative, while the other still remains positive. This is the reason why no stability regain occurs at point e'.

We will now assume continuously distributed delay, when the dynamic equation becomes

$$\dot{a}^e(t) = ka^e(t)\left(a - a^E(t)\right) \tag{5.22}$$

where

$$a^E(t) = \int_0^t W(t-s, m, T)a^e(s)ds.$$

The linearized version becomes

$$\dot{a}_\varepsilon^e(t) = -(ka)a_\varepsilon^E(t)$$

and by substituting exponential solution form

$$a_\varepsilon^e(t) = e^{\lambda t}u$$

and letting $t \to \infty$ in the definition of $a_\varepsilon^E(t)$, we get the characteristic equation as

$$\lambda + (ka)\left(1 + \frac{\lambda T}{\bar{m}}\right)^{-(m+1)} = 0$$

where

$$\bar{m} = \begin{cases} 1 & \text{if } m = 0, \\ m & \text{if } m \geq 1. \end{cases}$$

This equation can be rewritten as a polynomial equation

$$\lambda\left(1 + \frac{\lambda T}{\bar{m}}\right)^{m+1} + ka = 0. \tag{5.23}$$

Notice this is the same as Eq. (2.66) examined earlier in Chap. 2, when $\gamma = ka$. So we can apply the results obtained in Proposition 2.16 for our case with very minor modifications. Without delay $T = 0$, $\lambda = -ka < 0$, so the steady state is asymptotically stable. Next small values of m are examined. If $m = 0$, then the steady state is locally asymptotically stable with all values of T, a and k.

If $m = 1$, then stability occurs if

$$T < \frac{2}{ka} = \frac{T_1^*}{ka}$$

which is a hyperbola in ka. At $T = T_1^*/ka$, Hopf bifurcation occurs with the stability loss. At $m = 2$, the stability condition is

$$T < \frac{16}{9ka} = \frac{T_2^*}{ka} \simeq \frac{1.7778}{ka}$$

and at $m = 3$, the stability condition becomes

$$T < \frac{24(-7+5\sqrt{2})}{ka} = \frac{T_3^*}{ka} \simeq \frac{1.7056}{ka}.$$

Proposition 5.8 *The steady state with model* (5.22) *is locally asymptotically stable without delay and for $m = 0$ with all positive values of T, a and k. If $m \geq 1$, then local asymptotical stability occurs if*

$$T < \frac{T_m^*}{ka}$$

where $T_1^* = 2$, $T_2^* \simeq 1.7778$ and $T_3^* \simeq 1.7056$.

We can also modify dynamics (5.14) to the case of continuously distributed delays,

$$\dot{a}^e(t) = ka^e(t)\left[a - \left(\omega a^e(t) + (1-\omega)a^E(t)\right)\right]. \tag{5.24}$$

The linearized equation becomes

$$\dot{a}_\varepsilon^e(t) = -ka\omega a_\varepsilon^e(t) - ka(1-\omega)a_\varepsilon^E(t).$$

As before, assuming

$$a_\varepsilon^e(t) = e^{\lambda t}u$$

as the solution and letting $t \to \infty$ in the definition of $a^E(t)$, we have

$$(\lambda + ka\omega) + ka(1-\omega)\left(1 + \frac{\lambda T}{\bar{m}}\right)^{-(m+1)} = 0$$

or in polynomial form

$$(\lambda + ka\omega)\left(1 + \frac{\lambda T}{\bar{m}}\right)^{m+1} + ka(1-\omega) = 0. \tag{5.25}$$

Notice first that $\lambda \neq 0$, and if $\omega \geq 1/2$, then with $\text{Re}\lambda \geq 0$,

$$\left| (\lambda + ka\omega) \left(1 + \frac{\lambda T}{\bar{m}} \right)^{m+1} \right| > ka\omega \geq ka(1-\omega)$$

implying the local asymptotical stability of the steady state which means successful learning. So we assume next that $\omega < 1/2$. Equation (5.25) identical with (2.77) when γ is replaced by ka and ω by $1-\omega$. Therefore, all results concerning Eq. (2.77) remain valid in our case.

If $m = 0$, then the steady state is locally asymptotically stable with all parameter selections, $ak > 0$ and $0 \leq \omega \leq 1$. In the case of $m = 1$, we have two cases. If $1/9 < \omega < 1/2$, then the steady state is always locally asymptotically stable. As $\omega = 1/9$, the steady state is locally asymptotically stable if $Tka \neq 9$, and if $0 \leq \omega < 1/9$, then stability occurs if either

$$0 < Tka < (Tka)_A \text{ or } Tka > (Tka)_B$$

where

$$(Tka)_{A,B} = \frac{1 - 5\omega \pm \sqrt{(1-\omega)(1-9\omega)}}{4\omega^2}. \tag{5.26}$$

Assume next that $m = 2$; then, the stability condition is the following, which is equivalent to (2.80):

$$\boldsymbol{H}_3 = \alpha_0 \,(Tka)^3 + \alpha_1 \,(Tka)^2 + \alpha_2 \,(Tka) + \alpha_3 > 0 \tag{5.27}$$

where

$$\alpha_0 = (1-9\omega)\omega^2, \ \alpha_1 = 12\omega(1-5\omega), \ \alpha_2 = -12(11\omega - 3) \text{ and } \alpha_3 = -64.$$

The locus of $\boldsymbol{H}_3 = 0$ in the (ω, Tka) plane is shown in Fig. 5.4. The shaded area shows the instability region.

Proposition 5.9 *The steady state with model* (5.24) *is locally asymptotically stable with all positive values of* T, k *and* a *if* $\omega \geq \frac{1}{2}$, *and local asymptotic stability occurs in the following cases:*

(i) $m = 0$ *with all positive* T, k *and* a.
(ii) $m = 1$ *if* $\frac{1}{9} < \omega < \frac{1}{2}$ *with all positive* T, k *and* a;
if $\omega = \frac{1}{9}$ *with* $Tka \neq 9$;
if $0 \leq \omega < \frac{1}{9}$ *with* $0 < Tka < (Tka)_A$ *or* $Tka > (Tka)_B$
(iii) $m = 2$ *if* (5.27) *holds*

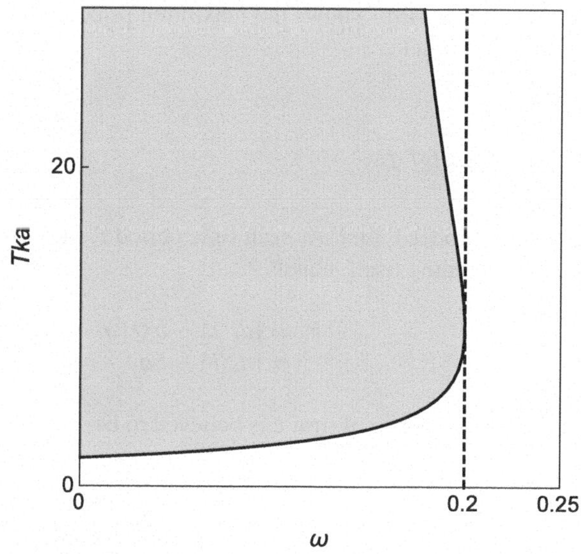

Fig. 5.4 Instability region in case of m = 2 is colored in gray

where $(Tka)_{A,B}$ *are given in* (5.26).

5.2 Learning in Oligopolies

An n-firm oligopoly is considered now, and for the sake of mathematical simplicity, no product differentiation is assumed. The price function is assumed to be linear, $p(Q) = a - bQ$ where Q is the industry output. Let the output of firm k be denoted by q_k; then assuming linear cost function $c_k q_k + d_k$, the profit of firm k is given as

$$\pi_k(q_1, \ldots, q_n) = (a - bQ)\, q_k - (c_k q_k + d_k)\,. \tag{5.28}$$

Since d_k adds only a constant term to the profit of the firm which has no effect on the optimal decision of the firm, we can assume that $d_k = 0$ as in the previous chapters. The firm cannot have negative profit, so it is also assumed that $a > c_k$.

The firms do not know the exact price function meaning that they have only its estimate, and based on repeated market information, they might be able to update their estimates, and even in certain cases, the estimates converge to the true values resulting in successful learning. In this section, three particular models are examined.

Model 1. Each firm knows the marginal price $-b$, but can only estimate the maximum price a. At time t, its estimate by firm k is denoted by $a_k(t)$.

Model 2. Each firm knows the market saturation point a/b and estimates the marginal price $-b_k(t)$.

Model 3. Each firm knows the maximum price a and estimates the marginal price $-b_k(t)$.

5.2.1 Model 1

Consider Model 1 first. At each time period t, firm k believes that the profit of any firm ℓ, including itself, equals

$$\begin{aligned}\bar{\pi}_\ell &= [a_k(t) - bQ]\,q_\ell - c_\ell q_\ell \\ &= [a_k(t) - bq_\ell - bQ_\ell - c_\ell]\,q_\ell\end{aligned} \tag{5.29}$$

so the best response of firm ℓ is believed to be

$$\begin{aligned}q_\ell &= \frac{a_k(t) - bQ_\ell - c_\ell}{2b} \\ &= \frac{a_k(t) - bQ + bq_\ell - c_\ell}{2b}\end{aligned}$$

where Q_ℓ is the output of the rest of the industry from the point of view of firm ℓ. So this believed best response function can be rewritten as

$$q_\ell = \frac{a_k(t) - bQ - c_\ell}{b}. \tag{5.30}$$

So firm k believes that the industry output is

$$Q = \sum_{\ell=1}^{n} q_\ell = \frac{na_k(t) - nbQ - C}{b}$$

where $C = \sum_{\ell=1}^{n} c_\ell$, so the believed industry output is

$$\bar{Q} = \frac{na_k(t) - C}{b(n+1)} \tag{5.31}$$

with equilibrium price

$$\bar{p}_k(t) = a_k(t) - b\bar{Q} = \frac{a_k(t) + C}{n+1}. \tag{5.32}$$

Firm k also computes its believed equilibrium output level based on (5.30) and (5.31) as

$$\bar{q}_k = \frac{a_k(t) - b\bar{Q} - c_k}{b} = \frac{a_k(t) + C - (n+1)c_k}{b(n+1)}. \tag{5.33}$$

Since all firms think in the same way, the actual industry output becomes

$$Q(t) = \sum_{k=1}^{n} \bar{q}_k = \frac{1}{b(n+1)} \left(\sum_{k=1}^{n} a_k(t) - C \right) \tag{5.34}$$

with actual market price

$$p(t) = a - bQ(t) = a - \frac{1}{n+1} \left(\sum_{k=1}^{n} a_k(t) - C \right). \tag{5.35}$$

Each firm then compares the believed and actual market prices. If the believed price is higher than the actual price, then firm k wants to decrease its belief by decreasing the value of $a_k(t)$. If the believed price is lower, then the firm wants to increase $a_k(t)$, and if the two prices are equal, then the firm keeps its belief. By assuming discrete time scales, this mechanism can be mathematically described by difference equation

$$\begin{aligned} a_k(t+1) &= a_k(t) + K_k \left[p(t) - \bar{p}_k(t) \right] \\ &= a_k(t) + K_k \left[a - \frac{1}{n+1} \left(\sum_{\ell=1}^{n} a_\ell(t) + a_k(t) \right) \right] \quad (k = 1, 2, \ldots, n) \end{aligned} \tag{5.36}$$

where $K_k > 0$ is the speed of adjustment of firm k. The steady state of this system is clearly $\bar{a}_k = a$ for all k, since by (5.36) the $a_k(t)$ values have to be identical and equal to a. The system is linear, so local asymptotical stability implies global asymptotical stability. The coefficient matrix of the system can be written as follows:

$$\boldsymbol{J} = \begin{pmatrix} 1 - \frac{2K_1}{n+1} & -\frac{K_1}{n+1} & \cdots & -\frac{K_1}{n+1} \\ -\frac{K_2}{n+1} & 1 - \frac{2K_2}{n+1} & \cdots & -\frac{K_2}{n+1} \\ \vdots & \vdots & \cdots & \vdots \\ -\frac{K_n}{n+1} & -\frac{K_n}{n+1} & \cdots & 1 - \frac{2K_n}{n+1} \end{pmatrix}.$$

By introducing the notation

$$\boldsymbol{D} = \left(1 - \frac{K_1}{n+1} - \lambda, \ldots, 1 - \frac{K_n}{n+1} - \lambda \right),$$

$$\boldsymbol{a} = \left(-\frac{K_1}{n+1}, \ldots, -\frac{K_n}{n+1} \right)^T$$

and

$$b^T = (1, \dots, 1)$$

we see that the characteristic polynomial has the special form

$$\begin{aligned} \varphi(\lambda) &= \det\left(\boldsymbol{D} + \boldsymbol{a}\boldsymbol{b}^T\right) \\ &= \det(\boldsymbol{D}) \det\left(\boldsymbol{I} + \boldsymbol{D}^{-1}\boldsymbol{a}\boldsymbol{b}^T\right) \\ &= \prod_{k=1}^{n}\left(1 - \frac{K_k}{n+1} - \lambda\right)\left[1 - \sum_{k=1}^{n} \frac{\frac{K_k}{n+1}}{1 - \frac{K_k}{n+1} - \lambda}\right] \end{aligned} \tag{5.37}$$

where relation (4.34) is used. The roots

$$\lambda = 1 - \frac{K_k}{n+1}$$

are between -1 and $+1$ if $K_k < 2(n+1)$. The other roots are solutions of equation

$$g(\lambda) = 1 - \sum_{k=1}^{n} \frac{\frac{K_k}{n+1}}{1 - \frac{K_k}{n+1} - \lambda} = 0.$$

Notice that

$$\lim_{\lambda \to \pm\infty} g(\lambda) = 1,\ g'(\lambda) < 0$$

and all poles are between -1 and $+1$. This equation is equivalent with a polynomial equation of degree S, where S is the number of different K_k values. There are $S-1$ roots between the poles, and the graph of $g(\lambda)$ has the same shape as shown in Fig. 4.1. Therefore, the last root before the smallest pole is also between -1 and $+1$ if $g(-1) > 0$, which can be rewritten as

$$\sum_{k=1}^{n} \frac{\frac{K_k}{n+1}}{2 - \frac{K_k}{n+1}} < 1. \tag{5.38}$$

Proposition 5.10 *The steady state of system* (5.36) *is asymptotically stable if* $K_k < 2(n+1)$, *and relation* (5.38) *holds.*

Consider now the symmetric case, when $K_1 = \cdots = K_n = K$. Then, the conditions are simplified as $K < 2(n+1)$ and (5.38) becomes

$$\frac{nK}{n+1} < 2 - \frac{K}{n+1}$$

or

$$K < 2. \tag{5.39}$$

Assume next that there is a unit time period delay for the firms to obtain market price information. Then, Eq. (5.35) is modified as

$$p(t-1) = a - \frac{1}{n+1}\left(\sum_{k=1}^{n} a_k(t-1) - C\right).$$

Therefore, the dynamic equation is modified in the following way:

$$a_k(t+1) = a_k(t) + K_k\left[a - \frac{1}{n+1}\left(\sum_{\ell=1}^{n} a_\ell(t-1) + a_k(t)\right)\right]. \tag{5.40}$$

This can be rewritten as a $2n$-dimensional system

$$\begin{aligned} a_k(t+1) &= a_k(t) + K_k\left[a - \frac{1}{n+1}\left(\sum_{\ell=1}^{n} z_\ell(t) + a_k(t)\right)\right] \\ z_k(t+1) &= a_k(t) \; (k = 1, 2, \ldots, n) \end{aligned} \tag{5.41}$$

with coefficient matrix

$$J = \begin{pmatrix} 1 - \frac{K_1}{n+1} & \cdot & \cdot & -\frac{K_1}{n+1} & \cdot & -\frac{K_1}{n+1} \\ \cdot & \cdot & \cdot & & \cdot & \\ \cdot & \cdot & 1 - \frac{K_n}{n+1} & -\frac{K_n}{n+1} & \cdot & -\frac{K_n}{n+1} \\ 1 & \cdot & \cdot & 0 & \cdot & 0 \\ \cdot & \cdot & \cdot & \cdot & \cdot & \cdot \\ \cdot & \cdot & 1 & 0 & \cdot & 0 \end{pmatrix}.$$

The eigenvalue equation can be written as

$$\begin{aligned} &\left(1 - \frac{K_k}{n+1}\right) u_k - \frac{K_k}{n+1} \sum_{\ell=1}^{n} v_\ell = \lambda u_k \\ &u_k = \lambda v_k \qquad\qquad (k = 1, 2, \ldots, n), \end{aligned}$$

that is, for $k = 1, 2, \ldots, n$ we have

$$\left[\lambda^2 - \left(1 - \frac{K_k}{n+1}\right)\lambda\right] v_k + \frac{K_k}{n+1}\sum_{\ell=1}^{n} v_\ell = 0. \tag{5.42}$$

We can assume that

$$\lambda \neq 0 \text{ and } \lambda \neq 1 - \frac{K_k}{n+1}.$$

By assuming that $K_k < 2(n+1)$, these values are between -1 and $+1$. Then

$$v_k = -\frac{\dfrac{K_k}{n+1}}{\lambda^2 - \left(1 - \dfrac{K_k}{n+1}\right)\lambda} V$$

with $V = \sum_{\ell=1}^{n} v_\ell$. By adding this equation for all values of k, we see that either $V = 0$ or

$$g(\lambda) = \sum_{k=1}^{n} \frac{\frac{K_k}{n+1}}{1 - \dfrac{K_k}{n+1} - \lambda} = \lambda. \tag{5.43}$$

If $V = 0$, then either

$$\lambda = 0 \text{ or } \lambda = 1 - \frac{K_k}{n+1} \text{ or } v_k = 0.$$

The first two cases do not disturb stability. In the third case (5.43) still holds, when summation is done for all values of k such that $v_k \neq 0$. This equation is equivalent with a polynomial equation of degree $S + 1$, where S is the number of different K_k values. All poles are between -1 and $+1$. Since $g'(\lambda) > 0$, the graph of $g(\lambda)$ has the same shape as shown in Fig. 4.2. There are $S - 1$ real roots between the poles, but the curve might not have more intersections with the 45° line implying the possibility of a pair of complex roots as well. Therefore, we cannot give simple stability condition.

Proposition 5.11 *The steady state of system* (5.40) *is asymptotically stable if* $K_k < 2(n+1)$ *for all* k, *and all roots of equation* (5.43) *are inside the unit circle.*

In the symmetric case $K_1 = \cdots = K_n = K$, and Eq. (5.43) simplifies as

$$\frac{nK}{n+1} = \lambda\left(1 - \frac{K}{n+1}\right) - \lambda^2$$

or

$$\lambda^2 - \lambda\left(1 - \frac{K}{n+1}\right) + \frac{nK}{n+1} = 0.$$

The roots are inside the unit circle if and only if

$$\frac{nK}{n+1} < 1$$
$$1 - \frac{K}{n+1} + \frac{nK}{n+1} + 1 > 0$$
$$-1 + \frac{K}{n+1} + \frac{nK}{n+1} + 1 > 0$$

which can be summarized as

$$K < \frac{n+1}{n}.$$

Consider next continuous time scales. Then, the difference $a_k(t+1) - a_k(t)$ in system (5.36) is replaced by the derivative $\dot{a}_k(t)$ to get the differential equations

$$\dot{a}_k(t) = K_k \left[a - \frac{1}{n+1} \left(\sum_{\ell=1}^{n} a_\ell(t) + a_k(t) \right) \right] \quad (k = 1, 2, \ldots, n). \tag{5.44}$$

The Jacobian of this system is the following:

$$\boldsymbol{J} = \begin{pmatrix} -\frac{2K_1}{n+1} & -\frac{K_1}{n+1} & \cdots & -\frac{K_1}{n+1} \\ -\frac{K_2}{n+1} & -\frac{2K_2}{n+1} & \cdots & -\frac{K_2}{n+1} \\ \vdots & \vdots & \cdots & \vdots \\ -\frac{K_n}{n+1} & -\frac{K_n}{n+1} & \cdots & -\frac{2K_n}{n+1} \end{pmatrix}.$$

Notice that its characteristic polynomial can be written as

$$\begin{aligned} \varphi(\lambda) &= \det\left(\boldsymbol{D} + \boldsymbol{a}\boldsymbol{b}^T\right) \\ &= \det(\boldsymbol{D}) \det\left(\boldsymbol{I} + \boldsymbol{D}^{-1}\boldsymbol{a}\boldsymbol{b}^T\right) \end{aligned}$$

with

$$\boldsymbol{D} = diag\left(-\frac{K_1}{n+1} - \lambda, \ldots, -\frac{K_n}{n+1} - \lambda\right)$$

$$\boldsymbol{a} = \left(-\frac{K_1}{n+1}, \ldots, -\frac{K_n}{n+1}\right)^T$$

and

$$\boldsymbol{b}^T = (1, \ldots, 1).$$

So

$$\varphi(\lambda) = \prod_{k=1}^{n} \left(-\frac{K_k}{n+1} - \lambda\right) \left[1 + \sum_{k=1}^{n} \frac{\frac{K_k}{n+1}}{\frac{K_k}{n+1} + \lambda} \right] = 0. \tag{5.45}$$

The eigenvalues

$$\lambda = -\frac{K_k}{n+1}$$

are negative. The others are the solutions of equation

$$g(\lambda) = 1 + \sum_{k=1}^{n} \frac{\frac{K_k}{n+1}}{\frac{K_k}{n+1} + \lambda} = 0.$$

Notice that

$$\lim_{\lambda \to \pm\infty} g(\lambda) = 1 \text{ and } g'(\lambda) < 0,$$

so the graph of $g(\lambda)$ has the same shape as shown in Fig. 4.1. All poles are negative, the roots between the poles are also negative, and the additional root below the smallest pole is also negative.

Proposition 5.12 *The steady state of system* (5.44) *is always asymptotically stable.*

Assume next that there is a fixed delay $\tau > 0$ in obtaining the market price; then, system (5.44) is modified as

$$\dot{a}_k(t) = K_k \left[a - \frac{1}{n+1} \left(\sum_{\ell=1}^{n} a_\ell(t-\tau) + a_k(t) \right) \right] \quad (k = 1, 2, \ldots, n). \tag{5.46}$$

The associated homogeneous equation is

$$\dot{a}_{k\varepsilon}(t) + \frac{K_k}{n+1} \left(\sum_{\ell=1}^{n} a_{\ell\varepsilon}(t-\tau) + a_{k\varepsilon}(t) \right) = 0$$

and by substituting the exponential solution form

$$a_{k\varepsilon}(t) = e^{\lambda t} u_k$$

we have the following equation

$$\left(\lambda + \frac{K_k}{n+1} \right) u_k + \frac{K_k}{n+1} \sum_{\ell=1}^{n} e^{-\lambda\tau} u_\ell = 0 \quad (k = 1, 2, \ldots, n)$$

with coefficient matrix

$$\begin{pmatrix} \lambda + \frac{K_1}{n+1}(1 + e^{-\lambda\tau}) & \frac{K_1}{n+1} e^{-\lambda\tau} & \cdots & \frac{K_1}{n+1} e^{-\lambda\tau} \\ \frac{K_2}{n+1} e^{-\lambda\tau} & \lambda + \frac{K_2}{n+1}(1 + e^{-\lambda\tau}) & \cdots & \frac{K_2}{n+1} e^{-\lambda\tau} \\ \vdots & \vdots & \ddots & \vdots \\ \frac{K_n}{n+1} e^{-\lambda\tau} & \frac{K_n}{n+1} e^{-\lambda\tau} & \cdots & \lambda + \frac{K_n}{n+1}(1 + e^{-\lambda\tau}) \end{pmatrix}.$$

Let

$$D - \left(\lambda + \frac{K_1}{n+1}, \ldots, \lambda + \frac{K_n}{n+1}\right),$$

$$a = \left(\frac{K_1}{n+1}e^{-\lambda\tau}, \ldots, \frac{K_n}{n+1}e^{-\lambda\tau}\right)^T$$

and

$$b^T = (1, \ldots, 1),$$

then the characteristic polynomial has the form

$$\begin{aligned}\varphi(\lambda) &= \det\left(D + ab^T\right)\\ &= \det(D)\det\left(I + D^{-1}ab^T\right)\\ &= \prod_{k=1}^{n}\left(\lambda + \frac{K_k}{n+1}\right)\left[1 + \sum_{k=1}^{n} \frac{\frac{K_k}{n+1}e^{-\lambda\tau}}{\lambda + \frac{K_k}{n+1}}\right] = 0.\end{aligned} \tag{5.47}$$

Proposition 5.13 *The steady state of system* (5.46) *is asymptotically stable if all roots of equation*

$$\sum_{k=1}^{n} \frac{\frac{K_k}{n+1}e^{-\lambda\tau}}{\lambda + \frac{K_k}{n+1}} = -1$$

have negative real parts.

No simple stability conditions can be given in general. Therefore, we consider the symmetric case when $K_1 = \cdots = K_n = K$. The eigenvalues $\lambda = -K/(n+1)$ are negative. The others are solutions of equation

$$\lambda + \frac{K}{n+1} + \frac{nK}{n+1}e^{-\lambda\tau} = 0. \tag{5.48}$$

Without delay $\tau = 0$ and $\lambda = -K < 0$, the steady state is asymptotically stable. By increasing the value of τ from zero, stability might be lost when $\lambda = iv$ $(v > 0)$. Substituting this solution into (5.48) and separating the real and imaginary parts, we have

$$\frac{K}{n+1} + \frac{nK}{n+1}\cos v\tau = 0 \tag{5.49}$$

and

$$v - \frac{nK}{n+1}\sin v\tau = 0 \tag{5.50}$$

implying that

$$\frac{n^2K^2}{(n+1)^2} = v^2 + \frac{K^2}{(n+1)^2}$$

or

$$v^2 = \frac{K^2}{(n+1)^2}\left(n^2-1\right)$$

showing that

$$v = \frac{K}{n+1}\sqrt{n^2-1}.$$

From (5.49) and (5.50), we see that the critical values are

$$\tau_m^* = \frac{1}{v}\left[\cos^{-1}\left(-\frac{1}{n}\right) + 2m\pi\right] \text{ for } m = 0, 1, \ldots \tag{5.51}$$

The direction of the stability switches can be obtained by Hopf bifurcation. Consider the eigenvalue as the function of the bifurcation parameter τ: $\lambda = \lambda(\tau)$. Implicitly differentiating equation (5.48) with respect to τ, we have

$$\lambda'(\tau) + \frac{nK}{n+1}e^{-\lambda\tau}\left(-\lambda'(\tau)\tau - \lambda(\tau)\right) = 0$$

implying that

$$\lambda'(\tau) = \frac{\dfrac{nK}{n+1}e^{-\lambda\tau}\lambda(\tau)}{1 - \dfrac{nK}{n+1}e^{-\lambda\tau}\tau}$$

$$= \frac{-\lambda^2 - \dfrac{K}{n+1}\lambda(\tau)}{1 + \dfrac{K}{n+1}\tau + \lambda(\tau)\tau}$$

where Eq. (5.48) was also used. At $\lambda = iv$,

$$\lambda'(\tau) = \frac{v^2 - i\dfrac{K}{n+1}v}{1 + \dfrac{K}{n+1}\tau + iv\tau}\,\frac{1 + \dfrac{K}{n+1}\tau - iv\tau}{1 + \dfrac{K}{n+1}\tau - iv\tau}$$

with real part having the same sign as

$$v^2\left(1 + \frac{K}{n+1}\tau\right) - \frac{K}{n+1}v^2\tau = v^2 > 0.$$

Therefore, at each critical value at least one pair of eigenvalues changes the sign of its real part from negative to positive.

Proposition 5.14 *In the symmetric case, the steady state of system* (5.46) *is asymptotically stable if* $\tau < \tau_0^*$, *and stability is lost with Hopf bifurcation at* $\tau = \tau_0^*$ *and cannot be regained with larger values of* τ.

Assume next that each firm has continuously distributed delays in obtaining market price information. Then, dynamic equation (5.46) becomes

$$\dot{a}_k(t) = K_k \left[a - \frac{1}{n+1} \left(\int_0^t W(t-s, m_k, T_k) \sum_{i=1}^n a_i(s) ds + a_k(t) \right) \right], \tag{5.52}$$

when it is now assumed that the length of the delay depends on firm k and not on the individual competitors. It is easy to see similarly to the previously discussed cases that the characteristic equation has the form

$$\left(\lambda + \frac{K_k}{n+1} \right) u_k + \frac{K_k}{n+1} \left(1 + \frac{\lambda T_k}{\bar{m}_k} \right)^{-(m_k+1)} \sum_{i=1}^n u_i = 0$$

or

$$\left(\frac{n+1}{K_k} \lambda + 1 \right) \left(1 + \frac{\lambda T_k}{\bar{m}_k} \right)^{m_k+1} u_k + \sum_{i=1}^n u_i = 0 \tag{5.53}$$

where

$$\bar{m}_k = \begin{cases} 1 & \text{if } m_k = 0, \\ m_k & \text{if } m_k \geq 1. \end{cases}$$

It is very complicated to locate the roots of equation (5.53) in general, so special cases are considered in the symmetric case when

$$T_1 = \cdots = T_n = T, \; m_1 = \cdots = m_n = m \text{ and } K_1 = \cdots = K_n = K,$$

so $u_1 = \cdots = u_n$.

If there is no delay, then $T = 0$, in which case Eq. (5.53) becomes

$$\frac{n+1}{K} \lambda + 1 + n = 0$$

implying that $\lambda = -K$. The steady state is asymptotically stable. If $m = 0$, then a quadratic equation is obtained,

$$\left(\frac{n+1}{K} \lambda + 1 \right) (1 + \lambda T) + n = 0$$

or

$$\frac{n+1}{K}T\lambda^2 + \left(\frac{n+1}{K} + T\right)\lambda + (1+n) = 0.$$

Since all coefficients are positive, all roots have negative real parts implying the asymptotical stability of the steady state.

Assume next that $m = 1$. Then (5.53) is a cubic equation,

$$\left(\frac{n+1}{K}\lambda + 1\right)\left(1 + 2T\lambda + T^2\lambda^2\right) + n = 0.$$

Introducing the simplifying notation

$$\alpha = \frac{K}{n+1},$$

we have

$$(\lambda+\alpha)\left(1+2T\lambda+T^2\lambda^2\right)+\alpha n = T^2\lambda^3 + \left(2T+\alpha T^2\right)\lambda^2 + (1+2\alpha T)\lambda + \alpha(n+1) = 0.$$

Since all coefficients are positive, all roots have negative real parts if and only if

$$\left(2T + \alpha T^2\right)(1 + 2\alpha T) > \alpha(n+1)T^2$$

where Theorem B.1 of Appendix B is used. This is a quadratic inequality in αT,

$$2\,(\alpha T)^2 + (4-n)\,(\alpha T) + 2 > 0. \tag{5.54}$$

The discriminant of the left hand side is

$$(4-n)^2 - 16 = n(n-8)$$

which is negative as $n < 8$. Then, (5.54) holds for all $\alpha T > 0$. If $n = 8$, then (5.54) holds if $\alpha T \neq 1$. If $n > 8$, then there are two real roots,

$$(\alpha T)^*_{1,2} = \frac{n - 4 \pm \sqrt{n(n-8)}}{4} \text{ with } (\alpha T)^*_1 < (\alpha T)^*_2. \tag{5.55}$$

Both are positive and (5.54) holds if

$$\alpha T < (\alpha T)^*_1 \text{ or } (\alpha T)^*_2 < \alpha T,$$

when the steady state is asymptotically stable. It can be proved that at both values $\alpha T = (\alpha T)^*_1$ and $\alpha T = (\alpha T)^*_2$ Hopf bifurcation occurs. Both stability switching curves are hyperbolas in the (α, T) plane.

Proposition 5.15 *The steady state of the symmetric model* (5.52) *is asymptotically stable if*

(*a*) there is no delay or $m = 0$
(*b*) $m = 1$ and $n < 8$ with all parameter values
or $n = 8$ with $\alpha T \neq 1$
or $n > 8$ and either $\alpha T < (\alpha T)_1^*$ or $(\alpha T)_2^* < \alpha T$
where $(\alpha T)_{1,2}^*$ are given by relation (5.55).

5.2.2 Model 2

Consider now Model 2. Similarly to the previous model, we consider the situation from the point of view of firm k. It believes that the marginal price is $-b_k(t)$ with known values of the market saturation point a/b. Therefore, it believes that the profit of each firm is given as

$$\bar{\pi}_\ell(q_1, \dots, q_n) = b_k(t)\left(\frac{a}{b} - Q_\ell - q_\ell\right) q_\ell - c_\ell q_\ell \tag{5.56}$$

with best response,

$$q_\ell = \frac{a}{b} - Q - \frac{c_\ell}{b_k(t)},$$

so firm k thinks that the industry output is

$$Q = \frac{na}{b} - nQ - \frac{1}{b_k(t)} C,$$

where $C = \sum_{\ell=1}^n c_\ell$. Therefore, the believed industry output is

$$\bar{Q} = \frac{1}{n+1}\left(\frac{na}{b} - \frac{C}{b_k(t)}\right)$$

with believed market price at the equilibrium

$$\begin{aligned} \bar{p}_k(t) &= b_k(t)\left(\frac{a}{b} - \bar{Q}\right) \\ &= \frac{1}{n+1}\left(\frac{a}{b} b_k(t) + C\right). \end{aligned} \tag{5.57}$$

In reality, however, all firms produce believed equilibrium output levels, so the actual industry output becomes

$$Q(t) = \sum\nolimits_{\ell=1}^{n} \left[\frac{a}{b} - \frac{1}{n+1}\left(n\frac{a}{b} - \frac{C}{b_\ell(t)} \right) - \frac{c_\ell}{b_\ell(t)} \right]$$

with actual market price

$$\begin{aligned} p(t) &= a - bQ \\ &= a - \sum\nolimits_{\ell=1}^{n} \left[a - \frac{1}{n+1}\left(na - \frac{b}{b_\ell(t)} C \right) - \frac{bc_\ell}{b_\ell(t)} \right] \\ &= \frac{a}{n+1} + b\sum\nolimits_{\ell=1}^{n} \frac{c_\ell}{b_\ell(t)} - \frac{bC}{n+1}\sum\nolimits_{\ell=1}^{n} \frac{1}{b_\ell(t)}. \end{aligned}$$

The comparison of the actual and believed market prices leads to the adjustment process

$$b_k(t+1) = b_k(t) + K_k \left[p(t) - \bar{p}_k(t) \right],$$

since larger $b_k(t)$ values give higher price from (5.57). That is,

$$b_k(t+1) = b_k(t) + K_k \left[\frac{a}{n+1} + b\sum_{\ell=1}^{n} \left(\frac{c_\ell}{b_\ell(t)} - \frac{C}{n+1}\frac{1}{b_\ell(t)} \right) - \frac{1}{n+1}\left(\frac{a}{b} b_k(t) + C \right) \right]. \tag{5.58}$$

At the steady state, the $\bar{b}_k$ values have to be identical and the common value $\bar{b}$ satisfies equation

$$\frac{a}{n+1}\left(1 - \frac{\bar{b}}{b}\right) - \frac{C}{n+1}\left(1 - \frac{b}{\bar{b}}\right) = 0$$

which may occur only when $\bar{b} = b$, the true value. This difference equation is nonlinear in state variables $b_k(t)$, so its dynamic properties can be examined by linearization. Notice that at the steady state,

$$\begin{aligned} \frac{\partial \left(p(t) - \bar{p}_k(t) \right)}{\partial b_k(t)} &= -\frac{bc_k}{b^2} + \frac{bC}{(n+1)b^2} - \frac{a}{(n+1)b} \\ &= \frac{C - (n+1)c_k - a}{(n+1)b} \end{aligned}$$

and for $\ell \neq k$

$$\begin{aligned} \frac{\partial \left(p(t) - \bar{p}_k(t) \right)}{\partial b_\ell(t)} &= -\frac{bc_\ell}{b^2} + \frac{bC}{(n+1)b^2} \\ &= \frac{C - (n+1)c_\ell}{(n+1)b}, \end{aligned}$$

so the linearized equations have the form

$$b_{k\varepsilon}(t+1) = b_{k\varepsilon}(t) + K_k \left[\frac{C-(n+1)c_k - a}{(n+1)b} b_{k\varepsilon}(t) + \sum_{\ell \neq k} \frac{C-(n+1)c_\ell}{(n+1)b} b_{\ell\varepsilon}(t) \right] \tag{5.59}$$

with coefficient matrix

$$\begin{pmatrix} \alpha_1 & K_1\beta_2 & \cdots & K_1\beta_n \\ K_2\beta_1 & \alpha_2 & \cdots & K_2\beta_n \\ \cdot & \cdot & \cdots & \cdot \\ K_n\beta_1 & K_n\beta_2 & \cdots & \alpha_n \end{pmatrix} = \boldsymbol{D} + \boldsymbol{a}\boldsymbol{b}^T$$

where we use the notation

$$\alpha_k = 1 + K_k \frac{C-(n+1)c_k - a}{(n+1)b},$$

$$\beta_\ell = \frac{C-(n+1)c_\ell}{(n+1)b},$$

$$\boldsymbol{D} = diag\,(\alpha_1 - K_1\beta_1, \ldots, \alpha_n - K_n\beta_n)$$

$$\boldsymbol{a} = (K_1, \ldots, K_n)^T$$

and

$$\boldsymbol{b}^T = (\beta_1, \ldots, \beta_n).$$

The characteristic polynomial is therefore

$$\begin{aligned} \varphi(\lambda) &= \det\left(\boldsymbol{D} + \boldsymbol{a}\boldsymbol{b}^T - \lambda \boldsymbol{I}\right), \\ &= \det(\boldsymbol{D} - \lambda\boldsymbol{I}) \det\left(\boldsymbol{I} + (\boldsymbol{D} - \lambda\boldsymbol{I})^{-1} \boldsymbol{a}\boldsymbol{b}^T\right) \\ &= \prod_{k=1}^n (\alpha_k - K_k\beta_k - \lambda) \left[1 + \sum_{k=1}^n \frac{K_k\beta_k}{\alpha_k - K_k\beta_k - \lambda} \right] = 0. \end{aligned} \tag{5.60}$$

Consider first equation

$$0 = \alpha_k - K_k\beta_k - \lambda = 1 - \frac{K_k a}{(n+1)b} - \lambda,$$

the root of which is

$$\lambda = 1 - \frac{K_k a}{(n+1)b}$$

which is between -1 and $+1$ if

$$K_k < \frac{2(n+1)b}{a}. \tag{5.61}$$

The other roots are the solutions of equation

$$g(\lambda) = 1 + \sum_{k=1}^{n} \frac{K_k \beta_k}{\alpha_k - K_k \beta_k - \lambda} = 0.$$

The signs of β_k are indeterminate; however, if the marginal costs are close to each other, then $\beta_k < 0$ for all k. Then $g'(\lambda) < 0$, and all poles are between -1 and $+1$, so the graph of $g(\lambda)$ is similar to the one shown in Fig. 4.1. If S denotes the number of different K_k values, then this equation is equivalent with a polynomial of degree S. We have $S - 1$ roots between the poles; all are between -1 and $+1$. There is an additional root before the smallest pole, which is larger than -1 if $g(-1) > 0$, or

$$\sum_{k=1}^{n} \frac{K_k \beta_k}{2 - \frac{K_k a}{(n+1)b}} > -1. \tag{5.62}$$

Proposition 5.16 *The steady state of system* (5.58) *is locally asymptotically stable if* $(n+1)c_k > C$, *and* (5.61) *holds for all* k; *furthermore,* (5.62) *is satisfied.*

In the symmetric case with

$$c_1 = \cdots = c_n = c \text{ and } K_1 = \cdots = K_n = K,$$

the stability conditions are as follows. Relation (5.61) becomes

$$K < \frac{2(n+1)b}{a}$$

and since $\beta = -c/\left[(n+1)b\right]$, (5.62) is simplified as

$$K < \frac{2(n+1)b}{a + nc}, \tag{5.63}$$

so this is the stability condition since this is the stronger (i.e., $a + nc > a$).

Proposition 5.17 *In the symmetric case, the steady state of system* (5.52) *is asymptotically stable if* (5.63) *holds.*

Assume next that there is a unit time period delay in obtaining the actual market price for the firms. Then, the dynamic equation (5.58) is modified as follows:

$$b_k(t+1) = b_k(t) + K_k \left[\frac{a}{n+1} + b \sum_{\ell=1}^{n} \left(\frac{c_\ell}{b_\ell(t-1)} - \frac{C}{n+1} \frac{1}{b_\ell(t-1)} \right) - \frac{1}{n+1} \left(\frac{a}{b} b_k(t) + C \right) \right] \tag{5.64}$$

since in the expression of $p(t)$, the time t is replaced by $t-1$ in $b_\ell(t)$. The linearized equation now becomes

$$b_{k\varepsilon}(t+1) = b_{k\varepsilon}(t) + K_k \left[\sum_{\ell=1}^{n} \beta_\ell b_{\ell\varepsilon}(t-1) - \frac{a}{(n+1)\,b} b_{k\varepsilon}(t) \right]$$

which can be rewritten as a $2n$-dimensional system:

$$\begin{aligned} b_{k\varepsilon}(t+1) &= b_{k\varepsilon}(t)\left(1 - \frac{aK_k}{(n+1)\,b}\right) + K_k \textstyle\sum_{\ell=1}^{n} \beta_\ell z_\ell(t), \\ z_\ell(t+1) &= b_{\ell\varepsilon}(t) \end{aligned}$$

with Jacobian matrix

$$J = \begin{pmatrix} 1 - \frac{aK_1}{(n+1)\,b} & \cdot & \cdot & K_1\beta_1 & \cdot & K_1\beta_n \\ \cdot & \cdot & \cdot & \cdot & \cdot & \cdot \\ \cdot & \cdot & 1 - \frac{aK_n}{(n+1)\,b} & K_n\beta_1 & \cdot & K_n\beta_n \\ 1 & \cdot & \cdot & 0 & \cdot & 0 \\ \cdot & \cdot & \cdot & \cdot & \cdot & \cdot \\ \cdot & \cdot & 1 & 0 & \cdot & 0 \end{pmatrix}.$$

The eigenvalue equations have the forms

$$\left(1 - \frac{aK_k}{(n+1)\,b}\right) u_k + K_k \textstyle\sum_{\ell=1}^{n} \beta_\ell v_\ell = \lambda u_k$$

$$u_k = \lambda v_k \ (k = 1, 2, \ldots, n)$$

where the first half of the eigenvector components is denoted by u_k, while the second half by v_k. Substituting the second equation to the first, we have

$$\left[\lambda^2 - \left(1 - \frac{aK_k}{(n+1)\,b}\right)\lambda\right] v_k - K_k \sum_{\ell=1}^{n} \beta_\ell v_\ell = 0. \tag{5.65}$$

Let

$$V = \sum_{\ell=1}^{n} \beta_\ell v_\ell,$$

and assume that

$$\lambda \neq 0 \text{ and } \lambda \neq 1 - \frac{aK_k}{(n+1)\,b},$$

which are inside the unit circle if

$$K_k < \frac{2(n+1)b}{a}.$$

In addition,

$$v_k = \frac{K_k}{\lambda\left[\lambda - \left(1 - \frac{aK_k}{(n+1)\,b}\right)\right]} V.$$

By multiplying both sides by β_k and adding for all k,

$$V = \sum_{k=1}^{n} \frac{K_k \beta_k}{\lambda\left[\lambda - \left(1 - \frac{aK_k}{(n+1)\,b}\right)\right]} V.$$

If $V = 0$, then either $v_k = 0$ or $\lambda = 0$ or

$$\lambda = 1 - \frac{aK_k}{(n+1)\,b}$$

which is between -1 and $+1$, if

$$K_k < \frac{2(n+1)b}{a}, \tag{5.66}$$

otherwise

$$\sum_{k=1}^{n} \frac{K_k \beta_k}{\lambda - \left(1 - \frac{aK_k}{(n+1)\,b}\right)} = \lambda. \tag{5.67}$$

Proposition 5.18 *The steady state of system* (5.64) *is locally asymptotically stable if* (5.66) *holds for all k, and Eq.* (5.67) *has roots only inside the unit circle.*

In the symmetric case

$$c_1 = c_2 = \cdots = c_n = c \text{ and } K_1 = \cdots = K_n = K \text{ so } \beta_k = -\frac{c}{(n+1)b}$$

so Eq. (5.67) has the form

$$\lambda^2 - \left(1 - \frac{aK}{(n+1)b}\right)\lambda + \frac{nKc}{(n+1)b} = 0.$$

The roots are inside the unit circle if

$$\frac{nKc}{(n+1)b} < 1,$$

$$1 - \frac{aK}{(n+1)b} + \frac{nKc}{(n+1)b} + 1 > 0,$$

$$-1 + \frac{aK}{(n+1)b} + \frac{nKc}{(n+1)b} + 1 > 0.$$

So the stability conditions are as follows:

$$K < \frac{(n+1)b}{nc} \text{ and } K < \frac{2(n+1)b}{a-nc} \text{ if } a > nc.$$

Assume next continuous time scales and that the firms have a fixed delay τ in obtaining market price. Then, Eq. (5.64) is replaced by the following differential equation for $k = 1, 2, \ldots, n$:

$$\dot{b}_k(t) = K_k \left[\frac{a}{n+1} + b \sum_{\ell=1}^{n} \left(\frac{c_\ell}{b_\ell(t-\tau)} - \frac{C}{n+1} \frac{1}{b_\ell(t-\tau)} \right) - \frac{1}{n+1} \left(\frac{a}{b} b_k(t) + C \right) \right]. \tag{5.68}$$

The linearized equation has the form

$$\dot{b}_{k\varepsilon}(t) = K_k \left[\sum_{\ell=1}^{n} \beta_\ell b_{\ell\varepsilon}(t-\tau) - \frac{a}{(n+1)\,b} b_{k\varepsilon}(t) \right]. \tag{5.69}$$

Substitution of the exponential solution form $b_{k\varepsilon}(t) = e^{\lambda t} u_k$ gives the following form of the linearized equations

$$\left(\lambda + \frac{aK_k}{(n+1)b} \right) u_k - K_k \sum_{\ell=1}^{n} \beta_\ell e^{-\lambda\tau} u_\ell = 0.$$

So the characteristic polynomial is

$$\varphi(\lambda) = \begin{pmatrix} P_1(\lambda) & -K_1\beta_2 e^{-\lambda\tau} & \cdots & -K_1\beta_n e^{-\lambda\tau} \\ -K_2\beta_1 e^{-\lambda\tau} & P_2(\lambda) & \cdots & -K_2\beta_n e^{-\lambda\tau} \\ \cdot & \cdot & \cdots & \cdot \\ -K_n\beta_1 e^{-\lambda\tau} & -K_n\beta_2 e^{-\lambda\tau} & \cdots & P_n(\lambda) \end{pmatrix}$$

where

$$P_k(\lambda) = \lambda + K_k \left(\frac{a}{(n+1)b} - \beta_k e^{-\lambda\tau} \right).$$

So

$$\begin{aligned}\varphi(\lambda) &= \det\left(\boldsymbol{D} + \boldsymbol{a}\boldsymbol{b}^T\right), \\ &= \det(\boldsymbol{D})\det\left(\boldsymbol{I} + \boldsymbol{D}^{-1}\boldsymbol{a}\boldsymbol{b}^T\right) \\ &= \prod_{k=1}^{n}\left(\lambda + \frac{aK_k}{(n+1)b}\right)\left[1 - \sum_{k=1}^{n}\frac{K_k\beta_k e^{-\lambda\tau}}{\lambda + \dfrac{aK_k}{(n+1)b}}\right] = 0.\end{aligned} \tag{5.70}$$

where

$$\boldsymbol{D} = diag\left(\lambda + \frac{aK_1}{(n+1)b}, \ldots, \lambda + \frac{aK_n}{(n+1)b}\right)$$

$$\boldsymbol{a} = (-K_1, \ldots, -K_n)^T$$

and

$$\boldsymbol{b}^T = (\beta_1 e^{-\lambda\tau}, \ldots, \beta_n e^{-\lambda\tau}).$$

The eigenvalues

$$\lambda = -\frac{aK_k}{(n+1)b}$$

are negative. Without delay $\tau = 0$, so the other eigenvalues are roots of equation

$$g(\lambda) = 1 - \sum_{k=1}^{n}\frac{K_k\beta_k}{\lambda + \dfrac{aK_k}{(n+1)b}} = 0.$$

Notice that $\beta_k < 0$ if the marginal costs are close to each other since $C < (n+1)c_k$ for all k. Then $g'(\lambda) < 0$; all poles are negative, so the roots between the poles are also negative. The shape of $g(\lambda)$ is the same as shown in Fig. 4.1, so the additional root is also negative.

Proposition 5.19 *The steady state of model* (5.68) *without delay is always locally asymptotically stable.*

In the case of $\tau > 0$, no simple stability condition can be given.

Proposition 5.20 *The steady state of model* (5.68) *with $\tau > 0$ is locally asymptotically stable if the roots of equation*

$$\sum_{k=1}^{n}\frac{K_k\beta_k e^{-\lambda\tau}}{\lambda + \dfrac{aK_k}{(n+1)b}} = 1 \tag{5.71}$$

have negative real parts.

In the symmetric case

$$K_1 = \cdots = K_n = K \text{ and } \beta_1 = \cdots = \beta_n = -\frac{c}{(n+1)b}$$

so Eq. (5.71) is the following:

$$\lambda + \frac{aK}{(n+1)b} + \frac{ncK}{(n+1)b} e^{-\lambda\tau} = 0. \tag{5.72}$$

At $\tau = 0$, the system is asymptotically stable. With increasing value of τ, stability might be lost when $\lambda = iv$. Substituting this value into this equation and separating the real and imaginary parts, we have

$$\frac{ncK}{(n+1)b} \cos v\tau = -\frac{aK}{(n+1)b}$$

$$\frac{ncK}{(n+1)b} \sin v\tau = v$$

implying that

$$\left(\frac{ncK}{(n+1)b}\right)^2 = \left(\frac{aK}{(n+1)b}\right)^2 + v^2$$

so

$$v = \frac{K}{(n+1)b}\sqrt{(nc)^2 - a^2}.$$

If $nc \le a$, then no stability switch occurs; otherwise, the critical values of delays are

$$\tau_m^* = \frac{(n+1)b}{K\sqrt{(nc)^2 - a^2}}\left[\cos^{-1}\left(-\frac{a}{nc}\right) + 2m\pi\right] \text{ for } m = 0, 1, 2, \ldots$$

The direction of stability switching is obtained by selecting τ as the bifurcation parameter, considering λ as function of τ and implicitly differentiating equation (5.72) with respect to τ to have

$$\lambda'(\tau) + \frac{ncK}{(n+1)b} e^{-\lambda(\tau)\tau}\left(-\lambda'(\tau)\tau - \lambda(\tau)\right) = 0$$

which is the same as in case of Model 1 when $nK/(n+1)$ is now replaced by $ncK/[(n+1)b]$. So at τ_0^* stability is lost, and it cannot be regained later with larger values of τ.

Assume next continuously distributed delays. Then, Eq. (5.69) is modified as

$$\dot{b}_{k\varepsilon}(t) = K_k\left[\sum_{\ell=1}^{n} \beta_\ell \int_0^t W(t-s, m_k, T_k) b_{\ell\varepsilon}(s) ds - \frac{a}{(n+1)\,b} b_{k\varepsilon}(t)\right]. \tag{5.73}$$

The characteristic equation is obtained by equating the determinant of system

$$\left(\lambda + \frac{aK_k}{(n+1)b}\right) u_k - K_k \left(1 + \frac{\lambda T_k}{\bar{m}_k}\right)^{-(m_k+1)} \sum_{\ell=1}^{n} \beta_\ell u_\ell = 0 \tag{5.74}$$

with zero.

The symmetric case will be only discussed, because no general stability condition can be derived from this equation. So assume that

$$K_1 = \cdots = K_n = K,\ T_1 = \cdots = T_n = T \text{ and } m_1 = \cdots = m_n = m,$$

and then we have the polynomial equation

$$\left(\lambda + \frac{aK}{(n+1)b}\right)\left(1 + \frac{\lambda T}{\bar{m}}\right)^{m+1} + n\frac{cK}{(n+1)b} = 0 \tag{5.75}$$

since in the symmetric case

$$\beta_k = -\frac{c}{(n+1)b} \text{ for } k = 1, 2, \ldots, n$$

as seen before.

If $T = 0$, then $\lambda < 0$, so steady state is asymptotically stable without delay as we have seen it earlier. If $m = 0$, then we have

$$\left(\lambda + \frac{aK}{(n+1)b}\right)(1 + \lambda T) + \frac{ncK}{(n+1)b} = T\lambda^2 + \left(1 + \frac{aKT}{(n+1)b}\right)\lambda + \frac{K}{(n+1)b}(a + nc) = 0$$

the roots of which have negative real parts, since all coefficients are positive, so the steady state is locally asymptotically stable. Let $m = 1$, then a cubic equation is obtained

$$\left(\lambda + \frac{aK}{(n+1)b}\right)(1 + 2\lambda T + T^2\lambda^2) + \frac{ncK}{(n+1)b} = 0$$

or

$$T^2\lambda^3 + \left(2T + \frac{aKT^2}{(n+1)b}\right)\lambda^2 + \left(1 + \frac{2aKT}{(n+1)b}\right)\lambda + \frac{K}{(n+1)b}(a + nc) = 0.$$

All coefficients are positive, so the roots have negative real parts if and only if

$$\left(2T + \frac{aKT^2}{(n+1)b}\right)\left(1 + \frac{2aKT}{(n+1)b}\right) > \frac{KT^2(a + nc)}{(n+1)b}$$

(see Theorem B.1 of Appendix B). This relation can be rewritten as a quadratic inequality of KT, which can be solved in the same way as shown for Eq. (5.54). Therefore, the details are left as a simple exercise to the interested readers.

5.2.3 Model 3

We now turn to Model 3, where each firm knows the maximum price, a, and estimates the marginal price, $-b_k(t)$. Consider again firm k, who believes that the profit of each firm is given as

$$\begin{aligned}\bar{\pi}_\ell &= [a - b_k(t)Q]\,q_\ell - c_\ell q_\ell \\ &= [a - b_k(t)Q_\ell - b_k(t)q_\ell - c_\ell]\,q_\ell,\end{aligned} \tag{5.76}$$

so firm k believes that the best response of firm ℓ is as follows:

$$q_\ell = \frac{a - c_\ell - b_k(t)Q}{b_k(t)}. \tag{5.77}$$

So the believed industry output becomes

$$Q = \sum_{\ell=1}^{n} q_\ell = \frac{na - C - b_k(t)nQ}{b_k(t)}$$

so firm k thinks that

$$\bar{Q} = \frac{na - C}{(n+1)b_k(t)} \tag{5.78}$$

with believed equilibrium price

$$\bar{p}_k(t) = a - b_k(t)\bar{Q} = \frac{a + C}{n+1}.$$

In reality, each firm produces its own believed equilibrium output level, so the actual industry output is

$$\begin{aligned}Q(t) &= \sum\nolimits_{\ell=1}^{n} \left(\frac{a - c_\ell}{b_\ell(t)} - \frac{na - C}{(n+1)b_\ell(t)} \right) \\ &= \frac{1}{n+1}\left[(a + C)\sum\nolimits_{\ell=1}^{n} \frac{1}{b_\ell(t)} - (n+1)\sum\nolimits_{\ell=1}^{n} \frac{c_\ell}{b_\ell(t)} \right]\end{aligned}$$

by Eq. (5.77) applied to firm ℓ. The corresponding actual market price is

$$p(t) = a - bQ(t)$$
$$= a - \frac{b}{n+1}\left[(a+C)\sum\nolimits_{\ell=1}^{n}\frac{1}{b_\ell(t)} - (n+1)\sum\nolimits_{\ell=1}^{n}\frac{c_\ell}{b_\ell(t)}\right].$$

Since price decreases by increasing values of $b_k(t)$, the adjustment process is now modified as

$$b_k(t+1) = b_k(t) + K_k\left(\bar{p}_k(t) - p(t)\right). \tag{5.79}$$

The steady states are the solutions of equations

$$\bar{p}_k(t) - p(t) = 0 \tag{5.80}$$

and notice that the steady states are the same with continuous dynamics, and models with delays. Rewriting Eq. (5.80), we have

$$\frac{a+C}{n+1} - a + \frac{b}{n+1}\left[(a+C)\sum_{\ell=1}^{n}\frac{1}{b_\ell(t)} - (n+1)\sum_{\ell=1}^{n}\frac{c_\ell}{b_\ell(t)}\right] = 0 \text{ for } k = 1, 2, \ldots, n.$$

Notice that for all values of k we have the same equation implying that there are infinitely many steady states in all versions of Model 3. The true knowledge $\bar{b}_k = b$ is one of them. Therefore at any steady state, the firms do not notice discrepancy between the believed and observed market prices implying that there is no learning possibility with this model.

5.3 Supplementary Notes and Discussions

Learning in games is an important issue since in many cases there is uncertainty in the strategy sets of the competitors as well as in the payoff functions of all players. A comprehensive summary of the different learning schemes in game theory is presented by Fudenberg and Levine (1998). Nonlinear Cournot models with uncertainty are discussed in Leonard and Nishimura (1999), and the effect of misspecified price functions on the stability of Cournot equilibria is analyzed in Chiarella and Szidarovszky (2001, 2004), and special learning processes are discussed in Bischi et al. (2010). In this book, we are not dealing with statistical learning; we rather focus on adaptive learning processes when the firms repeatedly update their estimates on certain parameters of the uncertain price function. In the case of asymptotically stable process, the estimates converge to their true values showing successful learning.

Section 5.1 is based on some ideas from the earlier Sect. 2.3 and on the papers Matsumoto and Szidarovszky (2014b, f, 2015f, g, 2016). Section 4.2 includes material from Matsumoto et al. (2012a; 2014) and models without delay from Bischi et al. (2010).

Nyarko (1991) solves the profit-maximizing problem without the knowledge of the slope and intersection of the linear demand function and demonstrates that using Bayesian updating might lead to cycles. Szidarovszky and Krawczyk (2004) introduces a simple adaptive learning scheme which becomes the basis for later works of the authors. Young (2004) discusses strategic learning, Cyert and DeGroot (1987) is devoted mainly to Bayesian analysis, and DeGroot (1970) is a nice summary of the methodology of Bayesian optimal statistical decisions what the firms can apply in cases of uncertainty and statistical learning.

Chapter 6
Oligopolies with Partial Cooperation

In the previous chapters on duopolies and oligopolies, it was assumed that the firms made their decisions independently of each other without any consideration about the competitors. In industries, however, the situation is much more complicated because there are several ways of forming joint ventures, having partial equity interests or being involved in indirect shareholdings which all make the profits of the firms dependent on the strategies and even on the profits of the others. This model is examined and the flip-flop dynamics of cartelizing groups is introduced.

Consider n firms with profit functions π_k $(k = 1, 2, \ldots, n)$ as before. We will next introduce some different models of shareholding interlocks and joint interests.

In the case of *joint ventures*, let $\alpha_{k\ell}$ denote the ownership interest of firm k in firm ℓ, then the payoff function of firm k is the following:

$$\bar{\pi}_k = \left(1 - \sum_{\ell \neq k} \alpha_{\ell k}\right) \pi_k + \sum_{\ell \neq k} \alpha_{k\ell} \pi_\ell, \tag{6.1}$$

where the multiplier of π_k is assumed to be positive. Notice first that

$$\sum_{k=1}^{n} \bar{\pi}_k = \sum_{k=1}^{n} \pi_k,$$

therefore model (6.1) represents profit sharing among the firms. By introducing the parameters

$$\varepsilon_{k\ell} = \frac{\alpha_{k\ell}}{1 - \sum_{i \neq k} \alpha_{ik}}$$

we can see that maximizing $\bar{\pi}_k$ is equivalent with maximizing

A. Matsumoto and F. Szidarovszky, *Dynamic Oligopolies with Time Delays*,
https://doi.org/10.1007/978-981-13-1786-6_6

$$\pi'_k = \pi_k + \sum_{\ell \neq k} \varepsilon_{k\ell}\pi_\ell. \tag{6.2}$$

In certain cases, the management of each firm considers its own interest in the other firms but ignores the interest of other firms in its own firm. This *partial equity interest* model is described directly by payoff functions (6.2).

If *indirect shareholding* is assumed, then the payoff of firm k can be given as

$$\bar{\pi}'_k = \pi_k + \sum_{\ell \neq k} \alpha_{k\ell}\bar{\pi}'_\ell. \tag{6.3}$$

when each firm considers its own profit and its return on equity holdings of the others. We can easily show that this model is also equivalent with (6.2). Introduce vectors

$$\bar{\mathbf{\Pi}}' = \left(\bar{\pi}'_k\right), \mathbf{\Pi} = (\pi_k) \text{ and matrix } \boldsymbol{D} = (\alpha_{k\ell}) \text{ with zero diagonal elements.}$$

Then, (6.3) can be rewritten as

$$\bar{\mathbf{\Pi}}' = \mathbf{\Pi} + \boldsymbol{D}\bar{\mathbf{\Pi}}'$$

implying that

$$\bar{\mathbf{\Pi}}' = (\boldsymbol{I} - \boldsymbol{D})^{-1}\,\mathbf{\Pi}. \tag{6.4}$$

The diagonal elements of $\boldsymbol{I} - \boldsymbol{D}$ are equal to unity, and since

$$\sum_{\ell \neq k} \alpha_{\ell k} < 1 \text{ for all } k,$$

matrix $\boldsymbol{I} - \boldsymbol{D}$ is an M-matrix with nonnegative inverse. Let $\delta_{k\ell}$ denote the elements of $(\boldsymbol{I} - \boldsymbol{D})^{-1}$, then (6.4) has the form

$$\bar{\pi}'_k = \sum_{\ell=1}^{n} \delta_{k\ell}\pi_\ell$$

being the same as (6.1), which is equivalent with (6.2).

Let π_k^G denote now the gross profit of firm k, then in the case of *net indirect shareholding*, the gross profits satisfy relations

$$\pi_k^G = \pi_k + \sum_{\ell \neq k} \alpha_{k\ell}\pi_\ell^G \tag{6.5}$$

and then the payoff of firm k becomes

$$\pi'_k = \left(1 - \sum_{\ell \neq k} \alpha_{\ell k}\right)\left(\pi_k + \sum_{\ell \neq k} \alpha_{k\ell}\pi_\ell^G\right). \tag{6.6}$$

From (6.5) similarly to (6.4), we have

$$\mathbf{\Pi} = (\boldsymbol{I} - \boldsymbol{D})\,\mathbf{\Pi}^G$$

with

$$\mathbf{\Pi}^G = \left(\pi_k^G\right)$$

and so from (6.6),

$$\mathbf{\Pi}' = \mathbf{\Delta}\left(\mathbf{\Pi} + \boldsymbol{D}\mathbf{\Pi}^G\right)$$

where

$$\mathbf{\Pi}' = \left(\pi_k'\right)$$

and $\mathbf{\Delta}$ is a diagonal matrix with diagonal elements

$$1 - \sum_{\ell \neq k} \alpha_{\ell k} \text{ for } k = 1, 2, \ldots, n.$$

So

$$\begin{aligned}\mathbf{\Pi}' &= \mathbf{\Delta}\left(\mathbf{\Pi} + \boldsymbol{D}\,(\boldsymbol{I} - \boldsymbol{D})^{-1}\,\mathbf{\Pi}\right)\\ &= \mathbf{\Delta}\,(\boldsymbol{I} - \boldsymbol{D})^{-1}\,\mathbf{\Pi},\end{aligned}$$

since

$$\begin{aligned}\boldsymbol{I} + \boldsymbol{D}\,(\boldsymbol{I} - \boldsymbol{D})^{-1} &= (\boldsymbol{I} - \boldsymbol{D})\,(\boldsymbol{I} - \boldsymbol{D})^{-1} + \boldsymbol{D}\,(\boldsymbol{I} - \boldsymbol{D})^{-1}\\ &= (\boldsymbol{I} - \boldsymbol{D} + \boldsymbol{D})\,(\boldsymbol{I} - \boldsymbol{D})^{-1}.\end{aligned}$$

Therefore, each component π_k' of vector $\mathbf{\Pi}'$ is a linear combination of the elements π_ℓ of vector $\mathbf{\Pi}$ leading to model (6.1), which is equivalent with (6.2).

Model (6.2) can be also interpreted by assuming that each firm takes certain parts of the profits of the others in addition to its own profit into account in its payoff function. This kind of behavior of the firms is known as *partial cooperation*. The coefficient $\varepsilon_{k\ell}$ is called the cooperation level of firm k toward firm ℓ. If $\varepsilon_{k\ell} = 0$ for all $\ell \neq k$, then the noncooperative oligopoly is obtained. If $\varepsilon_{k\ell} = 1$ for all $\ell \neq k$, then the firms have the common payoff maximizing their overall profit giving the basis for cooperative solution. Fair distributions are offered by the different cooperative game-theoretical solution concepts.

For the sake of mathematical simplicity, linear Cournot oligopolies will be considered, where each firm produces a single product which is related to the products of the competitors. It is assumed that the linear price of product k is given as it was defined in Chap. 4,

$$p_k = \alpha_k - q_k - \gamma \sum_{i \neq k} q_i \quad (1 \leq k \leq n) \tag{6.7}$$

where q_i is the production level (output) of firm i. Here, α_k is the maximum price and $\gamma \in [-1, 1]$ indicates the substitutability of the products. Zero fixed costs are assumed, since they have no effect on the optimal decisions of the firms. Assuming linear cost functions with marginal costs c_k, the profit of firm k is given as

$$\pi_k = q_k \left(\alpha_k - q_k - \gamma \sum_{i \neq k} q_i \right) - c_k q_k. \tag{6.8}$$

In order to avoid negative profits, we assume that for all k, $\alpha_k > c_k$ as in the previous chapters. Assuming partial cooperation among the firms as in (6.2), the payoff function of firm k becomes

$$\pi'_k = q_k \left(\alpha_k - q_k - \gamma \sum_{i \neq k} q_i - c_k \right) + \sum_{\ell \neq k} \varepsilon_{k\ell} q_\ell \left(\alpha_\ell - q_\ell - \gamma \sum_{i \neq \ell} q_i - c_\ell \right) \tag{6.9}$$

where we assume that $0 \le \varepsilon_{k\ell} < 1$ for all $\ell \neq k$; that is, each firm gives higher priority to its own profit than to the profits of other firms. Notice that

$$\frac{\partial \pi'_k}{\partial q_k} = \alpha_k - 2q_k - c_k - \gamma \sum_{\ell \neq k} q_\ell \left(1 + \varepsilon_{k\ell}\right)$$

and

$$\frac{\partial^2 \pi'_k}{\partial q_k^2} = -2 < 0$$

showing that π'_k is strictly concave in q_k, so the first-order conditions give maximum payoff values, as well as the best response function of firm k is given as

$$R_k\left(\boldsymbol{q}_{-k}\right) = \frac{\alpha_k - c_k - \gamma \sum_{\ell \neq k} q_\ell \left(1 + \varepsilon_{k\ell}\right)}{2} \tag{6.10}$$

where

$$\boldsymbol{q}_{-k} = \left(q_1, \ldots, q_{k-1}, q_{k+1}, \ldots, q_n\right).$$

By assuming interior best responses, the equilibrium output levels can be obtained by solving the linear equations

$$2q_k + \gamma \sum_{\ell \neq k} \left(1 + \varepsilon_{k\ell}\right) q_\ell = \alpha_k - c_k \tag{6.11}$$

for unknowns q_k $(1 \le k \le n)$. These equations can be solved analytically under the simplifying assumption that each firm treats the competitors equally: $\varepsilon_{k\ell} = \bar{\varepsilon}_k$ for all $\ell \neq k$. Then, (6.11) simplifies as

$$\alpha_k - c_k = 2q_k + \gamma(1+\bar{\varepsilon}_k)\sum_{\ell \neq k} q_\ell$$

$$= [2 - \gamma(1+\bar{\varepsilon}_k)]\, q_k + \gamma(1+\bar{\varepsilon}_k)\, Q$$

where $Q = \sum_{\ell=1}^{n} q_\ell$ is the industry output. So

$$q_k = -\frac{\gamma(1+\bar{\varepsilon}_k)}{2-\gamma(1+\bar{\varepsilon}_k)} Q + \frac{\alpha_k - c_k}{2-\gamma(1+\varepsilon_k)} \tag{6.12}$$

and by adding these equation for all values of k,

$$Q = -\sum_{k=1}^{n} \frac{\gamma(1+\bar{\varepsilon}_k)}{2-\gamma(1+\bar{\varepsilon}_k)} Q + \sum_{k=1}^{n} \frac{\alpha_k - c_k}{2-\gamma(1+\bar{\varepsilon}_k)}$$

from which we have

$$Q = \frac{\sum_{k=1}^{n} \dfrac{\alpha_k - c_k}{2-\gamma(1+\bar{\varepsilon}_k)}}{1 + \sum_{k=1}^{n} \dfrac{\gamma(1+\bar{\varepsilon}_k)}{2-\gamma(1+\bar{\varepsilon}_k)}}. \tag{6.13}$$

Notice first that $-1 \leq \gamma \leq 1$ and since $\bar{\varepsilon}_k < 1$ for all k, $\gamma(1+\bar{\varepsilon}_k) < 2$. The denominator can be rewritten as

$$1 + \sum_{k=1}^{n} \frac{-2+\gamma(1+\bar{\varepsilon}_k)+2}{2-\gamma(1+\bar{\varepsilon}_k)} = 1 - n + \sum_{k=1}^{n} \frac{2}{2-\gamma(1+\bar{\varepsilon}_k)}. \tag{6.14}$$

This is clearly positive if $\gamma \geq 0$ and its sign is indeterminate as $\gamma < 0$. Since the numerator of (6.13) is positive, we need (6.14) to be positive in order to guarantee that $Q > 0$.

Before we turn our attention to dynamics, we examine some interesting properties of this model. In the symmetric case without product differentiation,

$$\alpha_k = \alpha,\ c_k = c \text{ and } \gamma = 1.$$

Then from (6.13),

$$Q = \frac{(\alpha - c)\sum_{k=1}^{n} \dfrac{1}{1-\bar{\varepsilon}_k}}{1 - n + 2\sum_{k=1}^{n} \dfrac{1}{1-\bar{\varepsilon}_k}} = \frac{\alpha - c}{2 - \dfrac{n-1}{\sum_{k=1}^{n} \dfrac{1}{1-\bar{\varepsilon}_k}}} \tag{6.15}$$

from which we can see the following:

(a) With fixed values of $\bar{\varepsilon}_\ell$ $(\ell \neq k)$, the value of Q decreases with increasing value of $\bar{\varepsilon}_k$, meaning that increases in the cooperation levels decrease the industry output leading to higher price.
(b) If the difference $\alpha - c$ increases, then Q increases as well, so if the firms become "better qualified," then the industry output increases. The firms are called better qualified if they produce better product with same cost.
(c) Total cooperation is the case when $\bar{\varepsilon}_k = 1$ for all k. As all $\bar{\varepsilon}_k$ converge to 1, the second term in the denominator of (6.15) converges to zero, so

$$Q \to \frac{\alpha - c}{2}$$

which is the total profit-maximizing industry output, since the first-order condition of maximizing $Q(\alpha - Q - c)$ with respect to Q is $\alpha - 2Q - c = 0$ leading to the same solution.

6.1 Dynamic Models

Assuming discrete time scales and adjustment toward best responses, we get the dynamic equations

$$q_k(t+1) = q_k(t) + K_k \left[R_k \left(\boldsymbol{q}_{-k}(t) \right) - q_k(t) \right] \text{ for } k = 1, 2, \ldots, n \tag{6.16}$$

where $0 < K_k \leq 1$ is the speed of adjustment of firm k. This is a linear discrete system, so local asymptotical stability implies global asymptotical stability. Since for $\ell \neq k$,

$$\frac{\partial R_k \left(\boldsymbol{q}_{-k} \right)}{\partial q_\ell} = -\frac{\gamma((1 + \varepsilon_{k\ell}))}{2}.$$

The Jacobian has the form

$$\boldsymbol{J} = \begin{pmatrix} 1 - K_1 & K_1\gamma_{12} & \cdots & K_1\gamma_{1n} \\ K_2\gamma_{21} & 1 - K_2 & \cdots & K_2\gamma_{2n} \\ \cdot & \cdot & \cdots & \cdot \\ K_n\gamma_{n1} & K_n\gamma_{n2} & \cdots & 1 - K_n \end{pmatrix}$$

where

$$\gamma_{k\ell} = -\frac{\gamma(1 + \varepsilon_{k\ell})}{2}.$$

Clearly $-1 < \gamma_{k\ell} < 1$ for all $\ell \neq k$.

Proposition 6.1 *The equilibrium with model* (6.16) *is asymptotically stable if all eigenvalues of matrix* $\boldsymbol{J}$ *are inside the unit circle.*

Simple sufficient conditions can be obtained by assuming that any norm of matrix $\boldsymbol{J}$ is less than unity. In the case of the row norm, the condition is the following:

$$\sum_{\ell \neq k} |\gamma_{k\ell}| = \frac{|\gamma|}{2}\left(n - 1 + \sum_{\ell \neq k} \varepsilon_{k\ell}\right) < 1 \text{ for } k = 1, 2, \ldots, n.$$

In the case of the column norm, we have, similarly, that

$$\sum_{\ell \neq k} K_\ell \, |\gamma_{\ell k}| = \frac{|\gamma|}{2} \sum_{\ell \neq k} K_\ell \left(1 + \varepsilon_{\ell k}\right) < K_k \text{ for } k = 1, 2, \ldots, n.$$

Consider next the special case when $\bar{\varepsilon}_{k\ell} = \bar{\varepsilon}_k$; that is, each firm treats the others equally. Then

$$\gamma_{k\ell} = -\frac{\gamma\,(1 + \bar{\varepsilon}_k)}{2}$$

which will be now denoted by $\bar{\gamma}_k$. Matrix $\boldsymbol{J}$ has now the special form

$$\boldsymbol{J} = \boldsymbol{D} + \boldsymbol{a}\boldsymbol{b}^T$$

similarly to model (4.46), where

$$\boldsymbol{D} = diag\,(1 - K_1(1 + \bar{\gamma}_1), \ldots, 1 - K_n(1 + \bar{\gamma}_n))$$

$$\boldsymbol{a} = (K_1\bar{\gamma}_1, \ldots, K_n\bar{\gamma}_n)^T$$

and

$$\boldsymbol{b}^T = (1, \ldots, 1).$$

So the characteristic polynomial is given as

$$\begin{aligned}
\varphi(\lambda) &= \det\left(\boldsymbol{D} - \lambda \boldsymbol{I} + \boldsymbol{a}\boldsymbol{b}^T\right), \\
&= \det\left(\boldsymbol{D} - \lambda \boldsymbol{I}\right) \det\left(\boldsymbol{I} + \left(\boldsymbol{D} - \lambda \boldsymbol{I}\right)^{-1} \boldsymbol{a}\boldsymbol{b}^T\right) \\
&= \textstyle\prod_{k=1}^{n} \left(1 - K_k(1 + \bar{\gamma}_k) - \lambda\right) \left[1 + \sum_{k=1}^{n} \dfrac{K_k\bar{\gamma}_k}{1 - K_k(1 + \bar{\gamma}_k) - \lambda}\right] = 0.
\end{aligned} \tag{6.17}$$

The eigenvalues

$$\lambda = 1 - K_k\left(1 - \frac{\gamma\,(1 + \bar{\varepsilon}_k)}{2}\right)$$

are between -1 and $+1$ if for all k,

$$0 < K_k \left(1 - \frac{\gamma(1+\bar{\varepsilon}_k)}{2}\right) < 2. \tag{6.18}$$

Notice that $1 + \bar{\varepsilon}_k < 2$, so

$$-1 < \frac{\gamma(1+\bar{\varepsilon}_k)}{2} < 1$$

and therefore

$$0 < 1 - \frac{\gamma(1+\bar{\varepsilon}_k)}{2} < 2$$

implying that (6.18) holds for all $K_k \leq 1$. The other eigenvalues are the solutions of equation

$$g(\lambda) = 1 + \sum_{k=1}^{n} \frac{K_k\bar{\gamma}_k}{1 - K_k(1+\bar{\gamma}_k) - \lambda} = 0. \tag{6.19}$$

The poles of the left hand side are between -1 and $+1$. The sign of $\bar{\gamma}_k$ is the opposite of the sign of γ.

Assume first that $\gamma < 0$, thus $\bar{\gamma}_k > 0$ and $g'(\lambda) > 0$. The shape of (6.19) is the same as shown in Fig. 4.2. Equation (6.19) is equivalent with a polynomial equation of degree S, where S denotes the number of different $1 - K_k(1+\bar{\gamma}_k)$ values. There are $S - 1$ solutions between the poles, which are between -1 and $+1$, and there is an additional solution above the largest pole, which is less than 1 if $g(1) > 0$. That is,

$$1 - \sum_{k=1}^{n} \frac{\bar{\gamma}_k}{1+\bar{\gamma}_k} > 0. \tag{6.20}$$

If $\gamma > 0$, then $\bar{\gamma}_k < 0$ for all k, so $g'(\lambda) < 0$ and the graph of $g(\lambda)$ is the same as shown in Fig. 4.1. All solutions between the poles are between -1 and $+1$ and the additional solution below the smallest pole is greater than -1, if $g(-1) > 0$. That is,

$$1 + \sum_{k=1}^{n} \frac{K_k\bar{\gamma}_k}{2 - K_k(1+\bar{\gamma}_k)} > 0. \tag{6.21}$$

Assume next that there is a unit period time delay in obtaining information about the output of the other firms. Then, system (6.16) is modified as

$$q_k(t+1) = q_k(t) + K_k\left[R_k\left(\boldsymbol{q}_{-k}(t-1)\right) - q_k(t)\right] \text{ for } k = 1, 2, \ldots, n \tag{6.22}$$

which is rewritten as a $2n$-dimensional no-delay system

$$\begin{aligned} q_k(t+1) &= q_k(t) + K_k\left[R_k\left(z_{-k}(t-1)\right) - q_k(t)\right] \\ z_k(t+1) &= q_k(t) \end{aligned} \quad \text{for } k = 1, 2, \ldots, n \tag{6.23}$$

with Jacobian matrix

$$
J = \begin{pmatrix}
1-K_1 & & & 0 & K_1\gamma_{12} & \cdots & K_1\gamma_{1n} \\
& 1-K_2 & & K_2\gamma_{21} & 0 & \cdots & K_2\gamma_{2n} \\
& & \cdot & \cdot & \cdot & \cdots & \cdot \\
& & 1-K_n & K_n\gamma_{n1} & K_n\gamma_{n2} & \cdots & 0 \\
1 & & & 0 & 0 & \cdots & 0 \\
& 1 & & 0 & 0 & \cdots & 0 \\
& & \cdot & \cdot & \cdot & \cdots & \cdot \\
& & 1 & 0 & 0 & \cdots & 0
\end{pmatrix}.
$$

The eigenvalue equations have the form

$$(1-K_k)\,u_k + \sum\nolimits_{\ell \neq k} K_k \gamma_{k\ell} v_\ell = \lambda u_k$$

$$u_k = \lambda v_k$$

where the first n components of the eigenvector are u_k and the other components are v_k. That is,

$$\left[\lambda^2 - (1-K_k)\lambda\right] v_k - \sum_{\ell \neq k} K_k \gamma_{k\ell} v_\ell = 0. \tag{6.24}$$

Proposition 6.2 *Assume that for all k and ℓ,*

$$|\gamma_{k\ell}| < \frac{1}{n-1}.$$

Then, all eigenvalues are between -1 *and* $+1$ *implying the asymptotical stability of the equilibrium with model* (6.22).

Proof Notice that with $|v_k| = \max_{1 \leq j \leq n} \left|v_j\right|$, we have

$$\left|\sum_{\ell \neq k} K_k \gamma_{k\ell} v_\ell\right| < (n-1)\frac{1}{n-1} K_k\,|v_k| = K_k\,|v_k|\,,$$

so Eq. (6.24) can be rewritten as

$$\lambda^2 - (1-K_k)\lambda - K_k\Gamma_k = 0$$

with $|\Gamma_k| < 1$. Then, all conditions of (C9) of Appendix C hold:

$$
\begin{aligned}
-K_k\Gamma_k &< 1, \\
1 - K_k - K_k\Gamma_k + 1 &> 0, \\
-1 + K_k - K_k\Gamma_k + 1 &> 0,
\end{aligned}
$$

since $0 < K_k \leq 1$ and $|\Gamma_k| < 1$.

Since eigenvectors are nonzero, the determinant of the linear equation (6.24) is zero giving the characteristic equation

$$\varphi(\lambda) = \det \begin{pmatrix} P_1(\lambda) & -K_1\gamma_{12} & \cdots & -K_1\gamma_{1n} \\ -K_2\gamma_{21} & P_2(\lambda) & \cdots & -K_2\gamma_{2n} \\ \cdot & \cdot & \cdots & \cdot \\ -K_n\gamma_{n1} & -K_n\gamma_{n2} & \cdots & P_n(\lambda) \end{pmatrix} = 0 \tag{6.25}$$

where

$$P_k(\lambda) = \lambda^2 - (1 - K_k)\lambda.$$

Proposition 6.3 *The equilibrium with model* (6.22) *is asymptotically stable if and only if all roots of equation* (6.25) *are inside the unit circle.*

Consider next the special case, where $\gamma_{k\ell} = \bar{\gamma}_k$ for $\ell \neq k$ in which case

$$\begin{aligned}\varphi(\lambda) &= \det\left(\boldsymbol{D} + \boldsymbol{a}\boldsymbol{b}^T\right), \\ &= \det(\boldsymbol{D})\det\left(\boldsymbol{I} + \boldsymbol{D}^{-1}\boldsymbol{a}\boldsymbol{b}^T\right) \\ &= \prod_{k=1}^n \left(\lambda^2 - (1-K_k)\lambda + K_k\bar{\gamma}_k\right)\left[1 - \sum_{k=1}^n \frac{K_k\bar{\gamma}_k}{\lambda^2 - (1-K_k)\lambda + K_k\bar{\gamma}_k}\right] = 0,\end{aligned}$$

where

$$\boldsymbol{D} = diag\left(\lambda^2 - (1-K_1)\lambda + K_1\bar{\gamma}_1, \ldots, \lambda^2 - (1-K_n)\lambda + K_n\bar{\gamma}_n\right)$$

$$\boldsymbol{a} = (-K_1\bar{\gamma}_1, \ldots, -K_n\bar{\gamma}_n)^T$$

and

$$\boldsymbol{b}^T = (1, \ldots, 1).$$

The roots of equation

$$\lambda^2 - (1-K_k)\lambda + K_k\bar{\gamma}_k = 0$$

are inside the unit circle, since $-1 < \bar{\gamma}_k < +1$, so the proof of Proposition 6.2 applies in this case. The other eigenvalues solve equation

$$g(\lambda) = 1 - \sum_{k=1}^n \frac{K_k\bar{\gamma}_k}{\lambda^2 - (1-K_k)\lambda + K_k\bar{\gamma}_k} = 0. \tag{6.26}$$

Since there is the possibility of complex roots, no simple condition can be given for stability.

Proposition 6.4 *The equilibrium with model* (6.22) *with* $\gamma_{k\ell} = \bar{\gamma}_k$ *for* $\ell \neq k$ *is asymptotically stable if and only if all roots of equation* (6.26) *are inside the unit circle.*

Consider next continuous time scales, where Eq. (6.16) is modified to the differential equation

$$\dot{q}_k(t) = K_k\left[R_k\left(\boldsymbol{q}_{-k}(t)\right) - q_k(t)\right] \text{ for } k = 1, 2, \ldots, n. \tag{6.27}$$

The Jacobian of this system is the following,

$$\boldsymbol{J} = \begin{pmatrix} -K_1 & K_1\gamma_{12} & \cdots & K_1\gamma_{1n} \\ K_2\gamma_{21} & -K_2 & \cdots & K_2\gamma_{2n} \\ \cdot & \cdot & \cdots & \cdot \\ K_n\gamma_{n1} & K_n\gamma_{n2} & \cdots & -K_n \end{pmatrix}$$

which is the same as that of system (4.90) discussed earlier in Chap. 4.

Proposition 6.5 *The equilibrium with model* (6.27) *is asymptotically stable if and only if all eigenvalues of matrix* $\boldsymbol{J}$ *are either negative real numbers or complex with negative real parts.*

Consider next the special case, where $\gamma_{k\ell} = \bar{\gamma}_k$ for all $\ell \neq k$. Then, $\boldsymbol{J}$ has the special structure $\boldsymbol{D} + \boldsymbol{a}\boldsymbol{b}^T$ with

$$\boldsymbol{D} = diag\left(-K_1(1+\bar{\gamma}_1), \ldots, -K_n(1+\bar{\gamma}_n)\right)$$

$$\boldsymbol{a} = (K_1\bar{\gamma}_1, \ldots, K_n\bar{\gamma}_n)^T$$

and

$$\boldsymbol{b}^T = (1, \ldots, 1)$$

implying that the characteristic equation is the following:

$$\begin{aligned} \varphi(\lambda) &= \det\left(\boldsymbol{D} - \lambda\boldsymbol{I} + \boldsymbol{a}\boldsymbol{b}^T\right), \\ &= \det\left(\boldsymbol{D} - \lambda\boldsymbol{I}\right)\det\left(\boldsymbol{I} + \left(\boldsymbol{D} - \lambda\boldsymbol{I}\right)^{-1}\boldsymbol{a}\boldsymbol{b}^T\right) \\ &= \prod_{k=1}^{n}\left(-K_k\left(1+\bar{\gamma}_k\right) - \lambda\right)\left[1 - \sum_{k=1}^{n}\frac{K_k\bar{\gamma}_k}{K_k\left(1+\bar{\gamma}_k\right)+\lambda}\right] = 0. \end{aligned} \tag{6.28}$$

The roots $\lambda = -K_k\left(1+\bar{\gamma}_k\right)$ are negative, since $-1 < \bar{\gamma}_k < 1$ as seen before. The other solutions are roots of equation

$$g(\lambda) = 1 - \sum_{k=1}^{n}\frac{K_k\bar{\gamma}_k}{K_k\left(1+\bar{\gamma}_k\right)+\lambda} = 0. \tag{6.29}$$

If S denotes the number of different $K_k\left(1+\bar{\gamma}_k\right)$ values, then this equation is equivalent with a polynomial equation of degree S. Since the sign of $\bar{\gamma}_k$ depends on the sign of γ, we consider two cases.

If $\gamma > 0$, then $\bar{\gamma}_k < 0$ for all k, and therefore $g'(\lambda) < 0$, so the graph of $g(\lambda)$ is similar to that shown in Fig. 4.1. There are $S - 1$ negative roots between the negative poles, and one more real root is below the smallest pole which is also negative. Therefore, the equilibrium is asymptotically stable.

Assume next that $\gamma < 0$, then $\bar{\gamma}_k > 0$ for all k, $g'(\lambda) > 0$, and the graph of $g(\lambda)$ has the same shape as the one shown in Fig. 4.2. There are $S - 1$ roots between the poles, and there is one more root above the largest pole which is negative if $g(0) > 0$. That is, the equilibrium is asymptotically stable if

$$\sum_{k=1}^{n} \frac{\bar{\gamma}_k}{1+\bar{\gamma}_k} < 1.$$

Assume next that there is a $\tau > 0$ fixed delay in obtaining data about the output levels of the competitors. Then, model (6.27) changes as follows:

$$\dot{q}_k(t) = K_k \left[R_k \left(\mathbf{q}_{-k}(t-\tau) \right) - q_k(t) \right] \text{ for } k = 1, 2, \ldots, n. \tag{6.30}$$

Similarly to the earlier discussed models, it is easy to see that the eigenvalue equation is similar to model (4.87),

$$\varphi(\lambda) = \begin{pmatrix} K_1 + \lambda & -K_1\gamma_{12}e^{-\lambda\tau} & \cdots & -K_1\gamma_{1n}e^{-\lambda\tau} \\ -K_2\gamma_{21}e^{-\lambda\tau} & K_2 + \lambda & \cdots & -K_2\gamma_{2n}e^{-\lambda\tau} \\ \cdot & \cdot & \cdots & \cdot \\ -K_n\gamma_{n1}e^{-\lambda\tau} & -K_n\gamma_{n2}e^{-\lambda\tau} & \cdots & K_n + \lambda \end{pmatrix} = 0. \tag{6.31}$$

No simple stability condition can be given in general.

Proposition 6.6 *The equilibrium with model* (6.30) *is asymptotically stable if all roots of equation* (6.31) *are real negative or complex with negative real parts.*

Consider again the special case when $\gamma_{k\ell} = \bar{\gamma}_k$ for all $\ell \neq k$. In the case of model (4.87), we proved that

$$\varphi(\lambda) = \prod_{k=1}^{n} \left(\lambda + K_k \left(1 + \bar{\gamma}_k e^{-\lambda\tau} \right) \right) \left[1 - \sum_{k=1}^{n} \frac{K_k \bar{\gamma}_k e^{-\lambda\tau}}{\lambda + K_k \left(1 + \bar{\gamma}_k e^{-\lambda\tau} \right)} \right]$$

which is identical with equation (4.88). It was shown there that the root of equation

$$\lambda + K_k \left(1 + \bar{\gamma}_k e^{-\lambda\tau} \right) = 0$$

without delay is negative, and there is no stability switch if $|\bar{\gamma}_k| \le 1$, which is the case now. The roots of equation

$$1 - \sum_{k=1}^{n} \frac{K_k \bar{\gamma}_k e^{-\lambda\tau}}{\lambda + K_k \left(1 + \bar{\gamma}_k e^{-\lambda\tau} \right)} = 0$$

cannot be located in general, so we consider the symmetric case when

$$K_1 = \cdots = K_n = K \text{ and } \bar{\gamma}_1 = \cdots = \bar{\gamma}_n = \bar{\gamma}.$$

In this special case, this equation becomes identical to (4.80), where $-\gamma/2$ is replaced by $\bar{\gamma}$:

$$\lambda + K - (n-1)K\bar{\gamma}e^{-\lambda\tau} = 0.$$

Without delay, the root is

$$\lambda = K\left[(n-1)\bar{\gamma} - 1\right]$$

which is negative if $\bar{\gamma} < 1/(n-1)$. So equilibrium is

$$\text{stable if } \bar{\gamma} < \frac{1}{n-1} \text{ and unstable if } \bar{\gamma} > \frac{1}{n-1}.$$

Increasing the value of τ from zero, stability switch might occur, when $\lambda = iv,\ v > 0$. It was shown that the critical value is

$$v^* = K\sqrt{(n-1)^2\bar{\gamma}^2 - 1},$$

so if $|\bar{\gamma}| \leq 1/(n-1)$, then no stability switch occurs. Thus, if $|\bar{\gamma}| < 1/(n-1)$, then the equilibrium remains asymptotically stable with all $\tau > 0$. It was also shown that otherwise at each critical value of τ,

$$\tau_n^* = \begin{cases} \dfrac{1}{v^*}\left[\cos^{-1}\left(\dfrac{1}{(n-1)\bar{\gamma}}\right) + 2n\pi\right] & \text{if } \bar{\gamma} < 0, \\[2ex] \dfrac{1}{v^*}\left[-\cos^{-1}\left(\dfrac{1}{(n-1)\bar{\gamma}}\right) + 2n\pi\right] & \text{if } \bar{\gamma} > 0, \end{cases}$$

at least one eigenvalue pair changes its real part to positive from negative, so stability regain cannot occur. If $\gamma > 1/(n-1)$, then the equilibrium is unstable without delay and remains unstable for all $\tau > 0$, and if $\bar{\gamma} < -1/(n-1)$, then stability is lost at $\tau = \tau_0^*$ and it cannot be regained with larger values of τ.

We finally examine the case when τ is a continuously distributed delay. Model (6.30) is now the following:

$$\dot{q}_k(t) = K_k\left[R_k\left(\boldsymbol{q}^e_{-k}(t)\right) - q_k(t)\right] \text{ for } k = 1, 2, \ldots, n, \tag{6.32}$$

where

$$\boldsymbol{q}^e_{-k}(t) = \left(q_1^e(t), \ldots, q_{k-1}^e(t), q_{k+1}^e(t), \ldots, q_n^e(t)\right)$$

with

$$q_i^e(t) = \int_0^t W(t-s, m_k, T_k) q_i(s) ds$$

where the weighting function is the same as in the previous chapters. The homogeneous equation becomes

$$\dot{q}_k(t) = K_k \left[\sum_{\ell \neq k} \gamma_{k\ell} \int_0^t W(t-s, m_k, T_k) q_\ell(s) ds - q_k(t) \right] \text{ for } k = 1, 2, \ldots, n.$$

Substituting the exponential solution form $q_k(t) = e^{\lambda t} u_k$ and letting $t \to \infty$, we have the following linear equations for the variables u_k:

$$(\lambda + K_k) u_k - K_k \sum_{\ell \neq k} \gamma_{k\ell} \left(1 + \frac{\lambda T_k}{\bar{m}_k}\right)^{-(m_k+1)} u_\ell = 0 \text{ for } k = 1, 2, \ldots, n,$$

where

$$\bar{m}_k = \begin{cases} 1 & \text{if } m_k = 0 \\ m_k & \text{if } m_k \geq 1. \end{cases}$$

It is assumed again that the length of the delay depends on firm k and not on the individual competitors. The determinant of this system provides the characteristic equation

$$\det \begin{pmatrix} \lambda + K_1 & -K_1\gamma_{12}\left(1+\frac{\lambda T_1}{\bar{m}_1}\right)^{-(m_1+1)} & \cdots & -K_1\gamma_{1n}\left(1+\frac{\lambda T_1}{\bar{m}_1}\right)^{-(m_1+1)} \\ -K_2\gamma_{21}\left(1+\frac{\lambda T_2}{\bar{m}_2}\right)^{-(m_2+1)} & \lambda + K_2 & \cdots & -K_2\gamma_{2n}\left(1+\frac{\lambda T_2}{\bar{m}_2}\right)^{-(m_2+1)} \\ \cdot & \cdot & \cdots & \cdot \\ -K_n\gamma_{n1}\left(1+\frac{\lambda T_n}{\bar{m}_n}\right)^{-(m_n+1)} & -K_n\gamma_{n2}\left(1+\frac{\lambda T_n}{\bar{m}_n}\right)^{-(m_n+1)} & \cdots & \lambda + K_n \end{pmatrix} = 0 \tag{6.33}$$

which is almost identical to the characteristic equation of model (4.120).

Proposition 6.7 *The equilibrium with model* (6.32) *is asymptotically stable if all roots of equation* (6.33) *are negative real or complex with negative real parts.*

Consider the special case next, where $\gamma_{k\ell} = \bar{\gamma}_k$ ($\ell \neq k$), and then Eq. (6.33) becomes almost identical to the characteristic equation of model (4.115), which was examined there in detail. Therefore, that discussion needs only minor modification, so it is not repeated here.

6.2 Cartelizing Groups and Antitrust Threshold

An n-firm Cournot oligopoly is now considered, when a group $C = \{1, 2, \ldots, m\}$ of firms form a coalition and the members of this group cooperate with each other with common cooperation level ε and have no cooperation with other firms $k \notin C$. First, we will determine the equilibrium of this mixed behavioral oligopoly. For the sake of mathematical simplicity, we assume that there is no product differentiation, meaning that $\alpha_k = \alpha$ for $k = 1, 2, \ldots, n$ and $\gamma = 1$. The common price is therefore $p(Q) = \alpha - Q$. However, the usual linear price function is given as $p(Q) = a - bQ$ with some $b > 0$. Selecting $b = 1$ does not restrict generality, since after dividing the profit of the firms by b, the multiplier of Q becomes unity for all firms without changing their optimal decisions. To avoid negative profits, it is assumed that $\alpha > c_k$ for all firms.

The payoff functions of firms $k \notin C$ are their own profits. With the notation $Q_k = \sum_{\ell \neq k} q_\ell$, the profit of firm k is given as

$$\pi_k = (\alpha - q_k - Q_k - c_k)\, q_k$$

implying that its best response function is

$$q_k = \frac{\alpha - c_k - Q_k}{2} = \frac{\alpha - c_k - Q + q_k}{2}$$

where Q is the industry output. Therefore

$$q_k = \alpha - c_k - Q. \tag{6.34}$$

The payoffs of firms $k \in C$ are as follows:

$$\pi_k = (\alpha - q_k - Q_k - c_k)\, q_k + \varepsilon \sum_{\ell \in C_k} (\alpha - q_k - Q_k - c_\ell)\, q_\ell$$

where $C_k = C \backslash \{k\}$, so with the notation $\bar{Q} = \sum_{k \in C} q_k$, the best responses of firms $k \in C$ are given as

$$q_k = \frac{\alpha - c_k - (Q - q_k) - \varepsilon \left(\bar{Q} - q_k\right)}{2}$$

implying that

$$q_k = \frac{\alpha - c_k - Q - \varepsilon \bar{Q}}{1 - \varepsilon}. \tag{6.35}$$

Adding (6.35) for all firms $k \in C$,

$$\bar{Q} = \frac{m\alpha - \sum_{\ell \in C} c_\ell - mQ - m\varepsilon\bar{Q}}{1-\varepsilon}$$

which can be rewritten as

$$mQ + (m\varepsilon + (1-\varepsilon))\,\bar{Q} = m\alpha - \sum_{\ell \in C} c_\ell. \tag{6.36}$$

Adding (6.34) for all firms $k \notin C$, we obtain

$$Q - \bar{Q} = (n-m)\alpha - \sum_{\ell \notin C} c_\ell - (n-m)Q$$

or

$$(n-m+1)\,Q - \bar{Q} = (n-m)\alpha - \sum_{\ell \notin C} c_\ell. \tag{6.37}$$

Substituting $\bar{Q}$ from this equation into (6.36), we have

$$mQ + (m\varepsilon + (1-\varepsilon))\left[(n-m+1)\,Q - (n-m)\alpha + \sum_{\ell \notin C} c_\ell\right] = m\alpha - \sum_{\ell \in C} c_\ell$$

showing that the industry output becomes

$$Q = \frac{\alpha\,[m + (n-m)\,(m\varepsilon + (1-\varepsilon))] - \sum_{\ell \in C} c_\ell - (m\varepsilon + (1-\varepsilon))\sum_{\ell \notin C} c_\ell}{n+1+\varepsilon\,(m-1)\,(n-m+1)}. \tag{6.38}$$

To further simplify the mathematical formulation, consider the semi-symmetric case, when the firms $k \notin C$ have identical marginal cost c_x and the identical marginal cost of firms $k \in C$ is c_y. Because of this symmetry, the firms $k \notin C$ have identical equilibrium outputs q_x and firms $k \in C$ also have identical equilibrium production levels q_y. From (6.34) and (6.35) , we can conclude that

$$q_x = \alpha - c_x - mq_y - (n-m)q_x$$

and

$$(1-\varepsilon)q_y = \alpha - c_y - mq_y - (n-m)q_x - \varepsilon mq_y$$

which present two equations for the unknowns q_x and q_y:

$$(n-m+1)q_x + mq_y = \alpha - c_x,$$

$$(n-m)q_x + (1-\varepsilon+m+\varepsilon m)q_y = \alpha - c_y,$$

giving the solutions

$$q_x = \frac{\alpha - c_x - (c_x - c_y)m + (\alpha - c_x)(m-1)\varepsilon}{n + 1 + \varepsilon\,(m-1)\,(n-m+1)} \tag{6.39}$$

and

$$q_y = \frac{\alpha - c_y + (n-m)(c_x - c_y)}{n + 1 + \varepsilon\,(m-1)\,(n-m+1)} \tag{6.40}$$

and from (6.38), the equilibrium industry output equals

$$Q = \frac{(\alpha - c_x)\,(n-m)\,(m\varepsilon + (1-\varepsilon)) + m\left(\alpha - c_y\right)}{n + 1 + \varepsilon\,(m-1)\,(n-m+1)}.$$

Simple differentiation shows that

$$\frac{\partial q_x}{\partial \varepsilon} = \frac{(m-1)m\left[\alpha - c_y + (n-m)(c_x - c_y)\right]}{\left[n + 1 + \varepsilon\,(m-1)\,(n-m+1)\right]^2}$$

$$\frac{\partial q_y}{\partial \varepsilon} = \frac{-(m-1)\,(n-m+1)\left[\alpha - c_y + (n-m)(c_x - c_y)\right]}{\left[n + 1 + \varepsilon\,(m-1)\,(n-m+1)\right]^2}$$

and

$$\frac{\partial Q}{\partial \varepsilon} = \frac{-m(m-1)\left[\alpha - c_y + (n-m)(c_x - c_y)\right]}{\left[n + 1 + \varepsilon\,(m-1)\,(n-m+1)\right]^2}.$$

Therefore, we notice that

$$sign\left(\frac{\partial q_x}{\partial \varepsilon}\right) = -sign\left(\frac{\partial q_y}{\partial \varepsilon}\right) = -sign\left(\frac{\partial Q}{\partial \varepsilon}\right) = sign\left[\alpha - c_y + (n-m)(c_x - c_y)\right],$$

which is assumed to be positive since $q_y > 0$, implying that an increase in the cooperation level of the cartelizing group increases the outputs of the noncooperating firms and decreases their own as well as the industry output. The decrease of the industry output increases the market price.

The US Justice Department and the Department of Commerce are continuously investigating anticompetitive behavior such as forming coalition and cartelizing. One measure which is often used is the Herfindahl–Hirschman index. It is defined as the sum of the squared market shares of the firms in an industry:

$$I_{HH} = \sum_{k=1}^{n} \left(\frac{q_k}{Q}\right)^2. \tag{6.41}$$

In our case

$$I_{HH} = \frac{(n-m)q_x^2 + mq_y^2}{\left[(n-m)q_x + mq_y\right]^2} = \frac{(n-m)\left(\frac{q_x}{q_y}\right)^2 + m}{\left[(n-m)\left(\frac{q_x}{q_y}\right) + m\right]^2}$$

where

$$\frac{q_x}{q_y} = \frac{(\alpha - c_x) - (c_x - c_y)m + (\alpha - c_x)(m-1)\varepsilon}{(\alpha - c_y) + (n-m)(c_x - c_y)}.$$

We can easily find the range of the I_{HH} index. Notice first that with the notation $x_k = q_k/Q$, we have $\sum_{k=1}^n x_k = 1$ and therefore

$$\sum_{k=1}^n x_k^2 \le \left(\sum_{k=1}^n x_k\right)^2 = 1$$

since there are no nonnegative cross products on the left hand side, so $I_{HH} \le 1$. Based on the Cauchy inequality

$$1 = \left(\sum_{k=1}^n x_k\right)^2 = \left(\sum_{k=1}^n 1 \cdot x_k\right)^2 \le \left(\sum_{k=1}^n 1^2\right)\left(\sum_{k=1}^n x_k^2\right) = n\sum_{k=1}^n x_k^2$$

showing that $I_{HH} \ge 1/n$. The largest unit value is obtained with a single monopoly in the market, and the smallest $1/n$ value is obtained in a symmetric noncooperative industry. Therefore, the government is checking for higher I_{HH} index values to detect violation. The cooperating firms also check the I_{HH} index, and when it reaches a certain threshold, then they stop cooperating until the index falls under another threshold when they start cooperating again. Here, the dynamic extension of this model is a flip-flop dynamics having two possible steady states: the noncooperative equilibrium if antitrust violation is detected at this point and the partially cooperative equilibrium if no violation is detected here.

In the noncooperative case, the discrete dynamic system has the form of (6.16) with $\varepsilon_{k\ell} = 0$ for all k and ℓ, so $\gamma_{k\ell} = -\gamma/2 = -1/2$. The stability condition is now (6.21), which has the simple form

$$\sum_{k=1}^n \frac{K_k}{4 - K_k} < 1. \tag{6.42}$$

In a partially cooperative case, the discrete dynamic system is as follows:

$$q_k(t+1) = q_k(t) + K_k \left(\frac{\alpha - c_k - \sum_{\ell \neq k} q_\ell - \varepsilon \sum_{i \in C, i \neq k} q_i}{2} - q_k(t) \right) \text{ for } k = 1, 2, \ldots, m$$

$$q_k(t+1) = q_k(t) + K_k \left(\frac{\alpha - c_k - \sum_{\ell \neq k} q_\ell}{2} - q_k(t) \right) \text{ for } k = m+1, \ldots, n \tag{6.43}$$

with Jacobian matrix

$$\boldsymbol{J} = \begin{pmatrix} 1-K_1 & -\frac{K_1(1+\varepsilon)}{2} & \cdots & -\frac{K_1(1+\varepsilon)}{2} & -\frac{K_1}{2} & -\frac{K_1}{2} & \cdots & -\frac{K_1}{2} \\ -\frac{K_2(1+\varepsilon)}{2} & 1-K_2 & \cdots & -\frac{K_2(1+\varepsilon)}{2} & -\frac{K_2}{2} & -\frac{K_2}{2} & \cdots & -\frac{K_2}{2} \\ \cdot & \cdot & \cdots & \cdot & \cdot & \cdot & \cdots & \cdot \\ -\frac{K_m(1+\varepsilon)}{2} & -\frac{K_m(1+\varepsilon)}{2} & \cdots & 1-K_m & -\frac{K_m}{2} & -\frac{K_m}{2} & \cdots & -\frac{K_m}{2} \\ -\frac{K_{m+1}}{2} & -\frac{K_{m+1}}{2} & \cdots & -\frac{K_{m+1}}{2} & 1-K_{m+1} & -\frac{K_{m+1}}{2} & \cdots & -\frac{K_{m+1}}{2} \\ -\frac{K_{m+2}}{2} & -\frac{K_{m+2}}{2} & \cdots & -\frac{K_{m+2}}{2} & -\frac{K_{m+2}}{2} & 1-K_{m+2} & \cdots & -\frac{K_{m+2}}{2} \\ \cdot & \cdot & \cdots & \cdot & \cdot & \cdot & \cdots & \cdot \\ -\frac{K_n}{2} & -\frac{K_n}{2} & \cdots & -\frac{K_n}{2} & -\frac{K_n}{2} & -\frac{K_n}{2} & \cdots & 1-K_n \end{pmatrix}.$$

The eigenvalue equation can be written as

$$(1-K_k)\,u_k - \frac{K_k(1+\varepsilon)}{2} \textstyle\sum_{\ell=1,\ \ell \neq k}^{m} u_\ell - \frac{K_k}{2} \sum_{\ell=m+1}^{n} u_\ell = \lambda u_k \ \text{ for } k = 1, 2, \ldots, m,$$
$$(1-K_k)\,u_k - \frac{K_k}{2} \textstyle\sum_{\ell=1,\ \ell \neq k}^{n} u_\ell = \lambda u_k \ \text{ for } k = m+1, \ldots, n. \tag{6.44}$$

We will next derive a rational equation for unknown λ. Introduce first the notation

$$U = \sum_{k=1}^{n} u_k \text{ and } \bar{U} = \sum_{k=1}^{m} u_k,$$

and then equations in (6.44) can be rewritten in the following way:

$$(1-K_k)\,u_k - \frac{K_k(1+\varepsilon)}{2}\left(\bar{U} - u_k\right) - \frac{K_k}{2}(U - \bar{U}) = \lambda u_k \ \text{ for } k = 1, 2, \ldots, m,$$

$$(1-K_k)\,u_k - \frac{K_k}{2}(U - u_k) = \lambda u_k \ \text{ for } k = m+1, \ldots, n,$$

implying that

$$\left(1-\frac{K_k(1-\varepsilon)}{2}-\lambda\right)u_k=\frac{K_k\varepsilon}{2}\bar{U}+\frac{K_k}{2}U \text{ for } k=1,2,\ldots,m,$$

$$\left(1-\frac{K_k}{2}-\lambda\right)u_k=\frac{K_k}{2}U \text{ for } k=m+1,\ldots,n.$$

The roots

$$\lambda=1-\frac{K_k(1-\varepsilon)}{2} \text{ and } \lambda=1-\frac{K_k}{2}$$

are between -1 and $+1$, so they cannot disturb stability, so we may assume that λ is different from these values. Then

$$u_k=\frac{K_k\varepsilon\bar{U}+K_kU}{2-K_k(1-\varepsilon)-2\lambda} \text{ for } k=1,2,\ldots,m, \tag{6.45}$$

and

$$u_k=\frac{K_kU}{2-K_k-2\lambda} \text{ for } k=m+1,\ldots,n. \tag{6.46}$$

By adding (6.45) for $k=1,2,\ldots,m$ and (6.46) for $k=m+1,\ldots,n$, we get two equations for unknowns U and $\bar{U}$:

$$\bar{U}=\sum_{k=1}^{m}\frac{K_k\varepsilon}{2-K_k(1-\varepsilon)-2\lambda}\bar{U}+\sum_{k=1}^{m}\frac{K_k}{2-K_k(1-\varepsilon)-2\lambda}U$$

and

$$U-\bar{U}=\sum_{k=m+1}^{n}\frac{K_k}{2-K_k-2\lambda}U.$$

Solving the first equation for $\bar{U}$ and substituting it into the second equation and noticing that $U\neq 0$, we have

$$\frac{\sum_{k=1}^{m}\frac{K_k}{2-K_k(1-\varepsilon)-2\lambda}}{1-\sum_{k=1}^{m}\frac{K_k\varepsilon}{2-K_k(1-\varepsilon)-2\lambda}}-\left(1-\sum_{k=m+1}^{n}\frac{K_k}{2-K_k-2\lambda}\right)=0$$

and by further simplification,

$$\sum_{k=1}^{m}\frac{K_k}{2-K_k(1-\varepsilon)-2\lambda}-\left(1-\sum_{k=m+1}^{n}\frac{K_k}{2-K_k-2\lambda}\right)\left(1-\sum_{k=1}^{m}\frac{K_k\varepsilon}{2-K_k(1-\varepsilon)-2\lambda}\right)=0. \tag{6.47}$$

This equation is equivalent with a polynomial equation of degree n. In general, we cannot give simple stability conditions; however, in the semi-symmetric case, it is possible. Assume therefore that

$$K_k = K\ (1 \le k \le m) \text{ and } K_k = \bar{K}\ (m+1 \le k \le n),$$

and then this equation becomes

$$\frac{mK}{2 - K(1-\varepsilon) - 2\lambda} - \left(1 - \frac{(n-m)\,\bar{K}}{2 - \bar{K} - 2\lambda}\right)\left(1 - \frac{mK\varepsilon}{2 - K(1-\varepsilon) - 2\lambda}\right) = 0 \tag{6.48}$$

which is a quadratic equation for λ:

$$\lambda^2 + a_1\lambda + a_2 = 0$$

where

$$a_1 = -\frac{1}{2}\left[4 - K\,(m+1+(m-1)\varepsilon) - \bar{K}(n+1-m)\right]$$

and

$$a_2 = \frac{1}{4}\left[4 - 2K\,(m+1+(m-1)\varepsilon) - 2\bar{K}(n+1-m) + K\bar{K}\,(n+1+\varepsilon(m-1)(n+1-m))\right].$$

The sufficient conditions for the roots being inside the unit circle are given by (C9) of Appendix C:

$$\begin{aligned} a_2 &< 1, \\ a_1 + a_2 + 1 &> 0, \\ -a_1 + a_2 + 1 &> 0. \end{aligned} \tag{6.49}$$

We will next examine these conditions in detail and start with the first conditions:

$$1 - a_2 = \frac{1}{4}\left[2K\,(m+1) + 2\bar{K}(n+1-m) - K\bar{K}\,(n+1) + K\varepsilon(m-1)(2 - \bar{K}\,(n+1-m))\right].$$

At $\varepsilon = 0$, this is positive, since

$$\begin{aligned} 1 - a_2 &= \frac{1}{4}\left[4 - (2-K)(2-\bar{K}) + 2(K-\bar{K})m + \bar{K}(2-K)n\right] \\ &> \frac{1}{4}\left[4 - (2-K)(2-\bar{K}) + 2(K-\bar{K})m + \bar{K}(2-K)m\right] \\ &= \frac{1}{4}\left[4 - (2-K)(2-\bar{K}) + K(2-\bar{K})m\right] > 0. \end{aligned}$$

The Hopf boundary can be obtained by solving $1 - a_2 = 0$ to get

$$n = \frac{2K\,(m+1+(m-1)\varepsilon) + 2\bar{K}(1-m) + K\bar{K}\left((m-1)^2\varepsilon - 1\right)}{\bar{K}\,[-2 + K\,(1+(m-1)\varepsilon)]}. \tag{6.50}$$

Let the right hand side be denoted by $\varphi_H(m, \varepsilon)$, then since $m < n$ in a mixed behavioral oligopoly, we have to check if $m < \varphi_H(m, \varepsilon)$. Notice that

$$m - \varphi_H(m, \varepsilon) = \frac{-2\bar{K} - K\,(m + 1 + (m - 1)\varepsilon)\,(2 - \bar{K})}{\bar{K}\,[-2 + K\,(1 + (m - 1)\varepsilon)]}.$$

The numerator is clearly negative, and the Hopf boundary is above the 45° line in the (m, n) space if the denominator is positive, which is the case if

$$m > \frac{2 - K(1 - \varepsilon)}{K\varepsilon} = m^*. \tag{6.51}$$

Notice that the derivative of $1 - a_2$ with respect to n equals

$$\frac{\bar{K}}{4}\,[2 - K\,(1 + (m - 1)\,\varepsilon)]\,,$$

so we have the following fact. In the case of (6.51), the Hopf boundary is above the 45° line and $1 - a_2 > 0$ if $m < n < \varphi_H(m, \varepsilon)$; otherwise, the Hopf boundary is under the 45° line and $1 - a_2 > 0$ if $n > \varphi_H(m, \varepsilon)$ which holds in the entire feasible region $(m, n \mid n > m)$.

In the second condition,

$$a_1 + a_2 + 1 = \frac{1}{4}\left[K\bar{K}\,(n + 1 + (m - 1)(n + 1 - m)\varepsilon)\right] > 0$$

so it always holds.

In the case of the third condition

$$-a_1 + a_2 + 1 = \frac{1}{4}\left[16 - 4\bar{K}\,(n + 1 - m) - 4K\,(m + 1) + K\bar{K}\,(n + 1) - K\varepsilon(m - 1)(4 - \bar{K}\,(n + 1 - m))\right].$$

So the flip boundary is obtained by solving $-a_1 + a_2 + 1 = 0$:

$$n = \frac{(4 - K)(4 - \bar{K}) + 4(\bar{K} - K)m - K\varepsilon(m - 1)(4 + \bar{K}(m - 1))}{\bar{K}(4 - K - K\varepsilon(m - 1))}. \tag{6.52}$$

Let $\varphi_F(m, \varepsilon)$ denote the right hand side. At $\varepsilon = 0$, the denominator is positive and

$$\varphi_F(m, 0) = \frac{(4 - K)(4 - \bar{K}) + 4(\bar{K} - K)m}{\bar{K}(4 - K)}$$

$$= \frac{4(\bar{K} - K)}{\bar{K}(4 - K)}m + \frac{4 - \bar{K}}{\bar{K}}.$$

Notice that the derivative of $-a_1 + a_2 + 1$ with respect to n equals

$$\frac{\bar{K}}{4}\left[-4 + K(1-\varepsilon) + Km\varepsilon\right]$$

which is negative at $\varepsilon = 0$, and since the other two conditions are satisfied, we have the following result.

Proposition 6.8 *In the semi-symmetric noncooperative case, the equilibrium is asymptotically stable if*

$$m < n < \frac{4(\bar{K} - K)}{\bar{K}(4 - K)}m + \frac{4 - \bar{K}}{\bar{K}}. \tag{6.53}$$

We turn our attention to the case of $\varepsilon > 0$. The $\varphi_F(m, \varepsilon)$ boundary is above the 45° line if

$$m - \varphi_F(m, \varepsilon) = \frac{(4 - \bar{K})\left[Km(1+\varepsilon) - (4 - K(1-\varepsilon))\right]}{\bar{K}\left[4 - K(1-\varepsilon) - Km\varepsilon\right]} < 0. \tag{6.54}$$

We have to consider two cases next.

(a) $4 - K(1-\varepsilon) - Km\varepsilon > 0$, which occurs if

$$m < \frac{4 - K(1-\varepsilon)}{K\varepsilon} = m_1^*.$$

Then, $-a_1 + a_2 + 1 > 0$ if $n < \varphi_F(m, \varepsilon)$. Notice that (6.54) holds if

$$m < \frac{4 - K(1-\varepsilon)}{K(1+\varepsilon)} = m_2^* \left(< m_1^*\right).$$

In this case, the equilibrium is asymptotically stable if $m < n < \varphi_F(m, \varepsilon)$. If $m_2^* \leq m < m_1^*$, then the $\varphi_F(m, \varepsilon)$ boundary is under the 45° line, so this region violates the condition $n < \varphi_F(m, \varepsilon)$, so the equilibrium is unstable.

(b) $4 - K(1-\varepsilon) - Km\varepsilon < 0$, which occurs if $m > m_1^*$. In this case, $-a_1 + a_2 + 1 > 0$ if $n > \varphi_F(m, \varepsilon)$. Relation (6.54) always holds since $m_2^* < m_1^*$.

We can summarize the above derivation as follows.

Proposition 6.9 *Consider the following conditions:*

$$\begin{aligned}
&(1)\ 1 - a_2 > 0 && \textit{if } m \leq m^* \\
& && \textit{or } m > m^* \textit{ and } m < n < \varphi_H(m, \varepsilon), \\
&(2)\ -a_1 + a_2 + 1 > 0 && \textit{if } m < m_2^* \textit{ and } m < n < \varphi_F(m, \varepsilon) \\
& && \textit{or } m > m_1^* \textit{ and } n > \varphi_F(m, \varepsilon).
\end{aligned}$$

The equilibrium is asymptotically stable if both conditions (1) and (2) are satisfied.

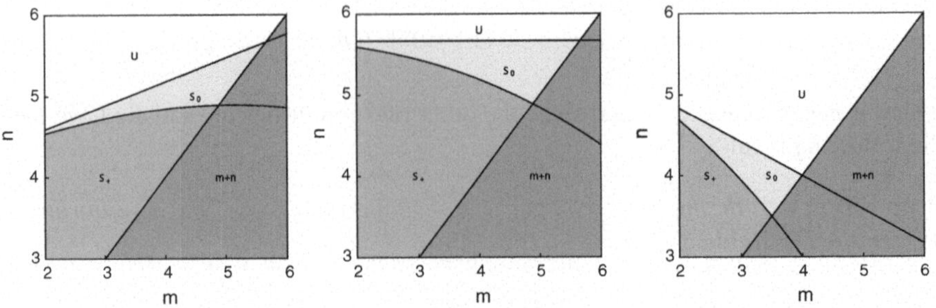

a. $\bar{K} = 0.8$ and $K = 0.6$ b. $\bar{K} = K = 0.6$ c. $\bar{K} = 0.6$ and $K = 0.8$

Fig. 6.1 Stability and instability regions

Simple differentiation shows that

$$\frac{\partial \varphi_F(m, \varepsilon)}{\partial \varepsilon} < 0$$

implying that the flip boundary moves downward with increasing values of ε.

Proposition 6.10 *The noncooperative semi-symmetric equilibrium is more stable than the partial cooperative case since the stability region is larger.*

Figure 6.1 shows the division of the (m, n) space with $m \geq 2$ and $n \geq 3$. The cooperative level $\varepsilon = 0.2$ is selected. The unstable white region is denoted by U, S_+ shows the stable region for the cooperative system, and $S_+ \cup S_0$ is the stability region in the noncooperative system. The 45° line $m = n$ is also shown in the figures as the steepest line. The locus of $\varphi_F(m, 0)$ is the less steep straight line, and the locus of $\varphi_F(m, \varepsilon)$ is the nonlinear curve in all figures.

Figure 6.1a is shown for $\bar{K} = 0.8$ and $K = 0.6$, Fig. 6.1b shows the case of $\bar{K} = K = 0.6$, and Fig. 6.1c is presented for $\bar{K} = 0.6$ and $K = 0.8$.

6.3 Supplementary Notes and Discussions

The concept of partial cooperation was introduced by Cyert and DeGroot (1973), which is equivalent with several versions of shareholding interlocks. Reynolds and Snapp (1982, 1986) introduced the models of joint ventures and partial equity interests among the firms. Indirect shareholding is examined in Flath (1991, 1992), while more recently Merlone (2007) introduced the model considering net indirect shareholding . Partial cooperation can be combined with the integration of anticompetitive behavior by the US Justice Department and by the Department of Commerce (Nwogugu 2005). For the sake of mathematical simplicity, only linear oligopolies were considered in this chapter.

Section 6.1 is based on mainly (Matsumoto et al. 2010a), and in the dynamic analysis, the ideas of the earlier Chap. 4 are used. Section 6.2 uses material from Matsumoto et al. (2008) with some additions.

Winston (2006) gives a comprehensive description of antitrust behavior in economics and explains the Herfindahl–Hirschman index (HHI) and its use in cases of horizontal merges . Collusions in other contexts are also examined by HHI (Porter and Zona 1999). Higher HHI value usually indicates higher price-cost margin (Viscusi et al. 2005). Partial cooperation and cartelizing groups are described in hyperbolic oligopolies in Matsumoto et al. (2010b). Some problems with applying the HHI in detecting cooperation of the firms are analyzed in Matsumoto et al. (2012b) when examples are presented illustrating that in cases of symmetric oligopolies the HHI does not depend on the cooperation level and in semi-symmetric cases there is the possibility that cooperation has a decreasing effect on the index.

Appendix A
Stability Switching Curves with Fixed Delays

Consider a general characteristic equation,

$$P_0(\lambda) + P_1(\lambda)e^{-\lambda\tau_1} + P_2(\lambda)e^{-\lambda\tau_2} + P_3(\lambda)e^{-\lambda(\tau_1+\tau_2)} = 0 \tag{A1}$$

where $P_0(\lambda),\ P_1(\lambda),\ P_2(\lambda)$ and $P_3(\lambda)$ are polynomials with real coefficients. The single-delay case is obtained when $P_2(\lambda) = P_3(\lambda) = 0$; the simple two-delay case is also a special case of (A1) with $P_3(\lambda) = 0$. The determination of stability switching curves is based on the following well-known result (Kuang 1993):

Lemma A.1 *Assume (τ_1, τ_2) varies continuously in $\mathbb{R}^2_+$. Then, the number of eigenvalues (with counting multiplicity) on $\mathbb{C}_+$ can change if an eigenvalue appears on or crosses the imaginary axis. Here,*

$$\mathbb{R}^2_+ = \{(\tau_1, \tau_2) \mid \tau_1, \tau_2 > 0\}$$

and

$$\mathbb{C}_+ = \{\lambda \mid \operatorname{Re} \lambda > 0\}.$$

The following assumptions are made in order to guarantee that (A1) is the characteristic equation of a delayed system and to exclude some trivial cases.

(i) There are finite number of eigenvalues on $\mathbb{C}_+$ under condition

$$\deg[P_0(\lambda)] \geq \max\{\deg[P_1(\lambda)],\ \deg[P_2(\lambda)],\ \deg[P_3(\lambda)]\}.$$

(ii) Zero frequency, $\lambda = 0$ is not an eigenvalue with any τ_1 and τ_2,

$$P_0(0) + P_1(0) + P_2(0) + P_3(0) \neq 0.$$

(iii) Polynomials $P_0(\lambda),\ P_1(\lambda),\ P_2(\lambda)$ and $P_3(\lambda)$ have no common zeros.

A. Matsumoto and F. Szidarovszky, *Dynamic Oligopolies with Time Delays*,
https://doi.org/10.1007/978-981-13-1786-6

(iv)

$$\lim_{\lambda\to\infty}\left[\left|\frac{P_1(\lambda)}{P_0(\lambda)}\right|+\left|\frac{P_2(\lambda)}{P_0(\lambda)}\right|+\left|\frac{P_3(\lambda)}{P_0(\lambda)}\right|\right]<1.$$

If (i) is violated, then Eq. (A1) can never be stable, since there are always eigenvalues with positive real parts (Bellman and Cooke 1963). If (ii) is violated, then 0 is always an eigenvalue, so (A1) cannot be stable. Assumption (iii) ensures that the characteristic equation has the lowest degree and is irreducible. For retarded-type delay equations, assumption (iv) is always satisfied (Bellman and Cooke 1963).

A.1 Single-delay Equations

Consider first equation

$$P_0(\lambda)+P_1(\lambda)e^{-\lambda\tau}=0 \tag{A2}$$

where $P_0(\lambda)$ and $P_1(\lambda)$ are polynomials of real coefficients. In this special case, Assumptions (i)–(iv) are simplified as follows:

(i) $\deg[P_0(\lambda)]\geq\deg[P_1(\lambda)]$.
(ii) $P_0(0)+P_1(0)\neq 0$.
(iii) Polynomials $P_0(\lambda)$ and $P_1(\lambda)$ have no common zeros.
(iv)

$$\lim_{\lambda\to\infty}\left|\frac{P_1(\lambda)}{P_0(\lambda)}\right|<1.$$

From Lemma A.1, we know that stability switch occurs when $\lambda=iv$ $(v>0)$ is an eigenvalue. Since eigenvalues occur in conjugate pairs, with assumption $v>0$, we do not lose generality. So

$$P_0(iv)+P_1(iv)e^{-iv\tau}=0. \tag{A3}$$

Since $\left|e^{-iv\tau}\right|=1$, we have

$$|P_0(iv)|=|P_1(iv)|.$$

From this equation, we can derive the set Ω of all crossing values of v. For each $v\in\Omega$, we have two possibilities.

If $P_0(iv)=P_1(iv)=0$, then $\tau>0$ is arbitrary. Otherwise,

$$e^{iv\tau}=-\frac{P_1(iv)}{P_0(iv)},$$

so

$$\tau_n=\frac{1}{v}\left[\arg\left(-\frac{P_1(iv)}{P_0(iv)}\right)+2n\pi\right] \tag{A4}$$

where n is an integer such that $\tau_n > 0$.

The direction of stability switch at each $\tau = \tau_n$ value can be determined by using Hopf bifurcation (Hale and Kocak 1991). Select τ as the bifurcation parameter, and consider the eigenvalue as the function of τ, $\lambda = \lambda(\tau)$. Differentiate implicitly equation (A2) with respect to τ:

$$P_0'\lambda'(\tau) + P_1'\lambda'(\tau)e^{-\lambda(\tau)\tau} + P_1 e^{-\lambda(\tau)\tau}\left(-\lambda'(\tau)\tau - \lambda(\tau)\right) = 0$$

implying that

$$\lambda'(\tau) = \frac{P_1 e^{-\lambda(\tau)\tau}\lambda(\tau)}{P_0' + \left(P_1' - P_1\tau\right)e^{-\lambda(\tau)\tau}} = \frac{-\lambda(\tau)P_0P_1}{P_0'P_1 - P_1'P_0 + P_0P_1\tau}$$

where we use Eq. (A2) to get

$$P_1 e^{-\lambda(\tau)\tau} = -P_0.$$

Therefore,

$$\frac{1}{\lambda'(\tau)} = \frac{-P_0'P_1 + P_1'P_0}{\lambda P_0 P_1} - \frac{\tau}{\lambda(\tau)},$$

and at $\lambda = iv$,

$$\mathrm{Re}\left[\frac{1}{\lambda'(\tau)}\right] = \mathrm{Re}\left[\frac{-P_0'}{\lambda(\tau)P_0} + \frac{P_1'}{\lambda(\tau)P_1}\right]$$
$$= \mathrm{Re}\left\{\frac{1}{iv}\left[-\mathrm{Re}\left(\frac{P_0'}{P_0}\right) - i\mathrm{Im}\left(\frac{P_0'}{P_0}\right) + \mathrm{Re}\left(\frac{P_1'}{P_1}\right) + i\mathrm{Im}\left(\frac{P_1'}{P_1}\right)\right]\right\}$$

having the same sign as

$$\mathrm{Im}\left[-\left(\frac{P_0'}{P_0}\right) + \left(\frac{P_1'}{P_1}\right)\right]. \tag{A5}$$

If this value is positive, then at the critical value $v \in \Omega$, the real part of an eigenvalue changes its sign from negative to positive meaning that the characteristic equation either remains unstable or loses stability. If the value (A5) is negative, then at least one pair of eigenvalues changes real parts from positive to negative, implying that the characteristic equation either remains unstable or becomes stable.

A.2 Two-delay Equations

Now, we assume that in Eq. (A1), $P_3(\lambda) = 0$, so the characteristic equation now has the form,

$$P_0(\lambda) + P_1(\lambda)e^{-\lambda\tau_1} + P_2(\lambda)e^{-\lambda\tau_2} = 0, \tag{A6}$$

and it is assumed that Assumptions (i)–(iv) also hold. That is,

(i) $\deg[P_0(\lambda)] \geq \max\{\deg[P_1(\lambda)], \deg[P_2(\lambda)]\}$.
(ii) $P_0(0) + P_1(0) + P_2(0) \neq 0$.
(iii) Polynomials $P_0(\lambda)$, $P_1(\lambda)$ and $P_2(\lambda)$ have no common zeros.
(iv)

$$\lim_{\lambda\to\infty}\left\{\left|\frac{P_1(\lambda)}{P_0(\lambda)}\right| + \left|\frac{P_2(\lambda)}{P_0(\lambda)}\right|\right\} < 1.$$

In the following discussion, we follow Gu et al. (2005). Divide both sides of (A6) by $P_0(\lambda)$ to get

$$1 + a_1(\lambda)e^{-\lambda\tau_1} + a_2(\lambda)e^{-\lambda\tau_2} = 0 \tag{A7}$$

where

$$a_1(\lambda) = \frac{P_1(\lambda)}{P_0(\lambda)} \text{ and } a_2(\lambda) = \frac{P_2(\lambda)}{P_0(\lambda)}.$$

However if $P_0(iv) = 0$ with some $v > 0$, then we might lose this root. In this case from (A6), we have

$$P_1(iv) + P_2(iv)e^{-iv(\tau_2-\tau_1)} = 0, \tag{A8}$$

so

$$e^{iv(\tau_2-\tau_1)} = -\frac{P_2(iv)}{P_1(iv)}$$

that is,

$$\tau_2^n - \tau_1^n = \frac{1}{v}\left[\arg\left(-\frac{P_2(iv)}{P_1(iv)}\right) + 2n\pi\right]. \tag{A9}$$

If $P_1(iv) = 0$ as well, but $P_2(iv) \neq 0$, then no solution exists; otherwise, if $P_1(iv) = P_2(iv) = 0$, then arbitrary positive τ_1 and τ_2 values are solutions.

Next, we assume that $P_0(iv) \neq 0$. If $a_1(iv) = 0$, then (A7) simplifies as

$$1 + a_2(iv)e^{-iv\tau_2} = 0$$

so τ_1 is arbitrary and

$$\tau_2^n = \frac{1}{v}\left[\arg\left(-a_2(iv)\right) + 2n\pi\right]. \tag{A10}$$

The case of $a_2(iv) = 0$ is similar.

If we consider the values of 1, $a_1(\lambda)e^{-\lambda\tau_1}$ and $a_2(\lambda)e^{-\lambda\tau_2}$ as complex vectors and we put them head to tail, then a triangle is obtained as shown in Fig. A.1. The sufficient and necessary conditions for these vectors to form a triangle are the following set of inequalities:

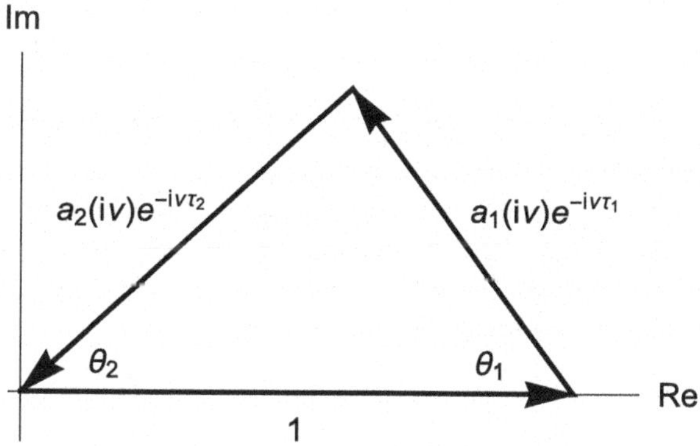

Fig. A.1 Triangle formed by the three vectors

$$|a_1(iv)| + |a_2(iv)| \geq 1, \tag{A11}$$

$$-1 \leq |a_1(iv)| - |a_2(iv)| \leq 1. \tag{A12}$$

The set Ω of all solutions of inequalities (A11)–(A12) gives the crossing set of the v values.

The law of cosine implies that the internal angles Θ_1 and Θ_2 can be obtained as

$$\Theta_1 = \cos^{-1}\left(\frac{1 + |a_1(iv)|^2 - |a_2(iv)|^2}{2\,|a_1(iv)|}\right) \tag{A13}$$

and

$$\Theta_2 = \cos^{-1}\left(\frac{1 + |a_2(iv)|^2 - |a_1(iv)|^2}{2\,|a_2(iv)|}\right) \tag{A14}$$

for all $v \in \Omega$ such that $a_1(iv)$ and $a_2(iv)$ differ from zero. Since the triangle can be formed above and under the real axis and the arguments of the three sides of the triangle are 0 and $\arg[a_1(iv)] - v\tau_1$ and $\arg[a_2(iv)] - v\tau_2$, we have

$$\tau_1^{n\pm}(v) = \frac{1}{v}\{\arg[a_1(iv)] + (2n-1)\pi \pm \Theta_1\} \tag{A15}$$

and

$$\tau_2^{m\pm}(v) = \frac{1}{v}\{\arg[a_2(iv)] + (2m-1)\pi \mp \Theta_2\} \tag{A16}$$

where n and m are integers such that τ_1 and τ_2 are positive.

Notice that inequalities (A11)–(A12) can be rewritten as polynomial inequalities with finitely many roots; therefore, Ω consists of finitely many intervals $\Omega = \cup_{k=1}^{N}\Omega_k$. Define

$$T_{n,m}^{k\pm} = \left\{\left(\tau_1^{n\pm}(v), \tau_2^{m\pm}(v)\right) \mid v \in \Omega_k\right\} \tag{A17}$$

and

$$T^k = \cup_{n=-\infty}^{\infty} \cup_{m=-\infty}^{\infty} \left(T_{n,m}^{k+} \cup T_{n,m}^{k-}\right) \cap \mathbb{R}_+^2. \tag{A18}$$

If we do not restrict $\arg[a_1(iv)]$ and $\arg[a_2(iv)]$ to be in interval $[0, 2\pi]$ but make it a continuous function of v in interval Ω_k, then with fixed n and m, $T_{n,m}^{k+}$ and $T_{n,m}^{k-}$ become continuous curves. The set T of all stability switching curves is the union of the curves of T^k. Notice that the left endpoint v_k^ℓ and the right endpoint v_k^r of Ω_k must satisfy at least one of inequalities (A11) or (A12) with equality. That is, one of the following inequalities must hold:

$$|a_1(iv)| + |a_2(iv)| = 1, \tag{A19}$$

$$|a_2(iv)| - |a_1(iv)| = 1, \tag{A20}$$

$$|a_1(iv)| - |a_2(iv)| = 1. \tag{A21}$$

The cases $v_k^\ell = 0$ and $v_k^r = \infty$ are also possible.

If (A19) holds, then $\Theta_1 = \Theta_2 = 0$ and $T_{n,m}^{k+}$ is connected with $T_{n,m}^{k-}$ at this end. If (A20) holds, then similarly $\Theta_1 = \pi$, $\Theta_2 = 0$, so $T_{n,m}^{k+}$ is connected with $T_{n+1,m}^{k-}$ at this end. If (A21) is satisfied, then $\Theta_1 = 0$, $\Theta_2 = \pi$ and $T_{n,m}^{k+}$ is connected with $T_{n,m-1}^{k-}$ at this end. If $v_k^\ell = 0$, then as $v \to 0$, both $T_{n,m}^{k+}$ and $T_{n,m}^{k-}$ converge to ∞ with asymptote

$$\frac{\tau_2^{m\pm}}{\tau_1^{n\pm}} \to \frac{\arg[a_2(0)] + (2m-1)\pi \mp \Theta_2(0)}{\arg[a_1(0)] + (2n-1)\pi \pm \Theta_1(0)}. \tag{A22}$$

Therefore for each k, the stability switching curve is the intersection of $\mathbb{R}_+^2$ and a series of curves belonging to one of the following types:

a series of closed curves;
a series of spiral-like curves with either horizontal, or vertical, or diagonal axes;
a series of open-ended curves whose both ends approach ∞.

Before discussing the direction of stability switches at a point of T^k, we make the following comments. The direction of the curve is called positive if it corresponds to increasing values of v. If the curve passes through an endpoint of Ω_k, then its direction is reversed. When we head in the positive direction of the curve, then the region on our left hand side is called the region on the left. The region on the right is similar. Let

$$R_\ell = \text{Re}\left[a_\ell(iv)e^{-i\tau_\ell v}\right] \text{ and } I_\ell = \text{Im}\left[a_\ell(iv)e^{-i\tau_\ell v}\right] \text{ for any } v \in \Omega \text{ and } \ell = 1, 2.$$

Gu et al. (2005) proved the following result:

Theorem A.2 *Let $v \in \Omega_k$ and $(\tau_1, \tau_2) \in T^k$ such that iv is a simple pure complex eigenvalue. Then, as point (τ_1, τ_2) moves from the right to the left of the corresponding* *curve of T^k, a pair of eigenvalues cross the imaginary axis to the right if $R_2 I_1 - R_1 I_2 > 0$. If the inequality is reversed, then the crossing is in the opposite direction.*

The application of the method derived above consists of the following steps:

Step 1. Solve equations $P_0(iv) = 0$, $P_1(iv) = 0$ and $P_2(iv) = 0$, and with all solutions $v > 0$, solve the corresponding single-delay equation.
Step 2. Solve inequalities (A11)–(A12) to find the crossing set of the v values. Notice that Ω consists of finitely many intervals Ω_k $(k = 1, 2, \ldots, N)$.
Step 3. Use equations (A13)–(A14) to find Θ_1 and Θ_2 as functions of v.
Step 4. Use equations (A15)–(A16) to compute the curves $(\tau_1^{n\pm}, \tau_2^{m\pm})$ for all Ω_k.
Step 5. Find the directions of stability switches based on Theorem A.2.

A.3 Three-delay Equations

Now, we turn our attention to Eq. (A1), where the three delays are not independent; one delay is the sum of the other two. In the following discussion, we follow Lin and Wang (2012). With $\lambda = iv$, Eq. (A1) can be rewritten as

$$P_0(iv) + P_1(iv)e^{-iv\tau_1} + \left[P_2(iv) + P_3(iv)e^{-iv\tau_1}\right] e^{-iv\tau_2} = 0 \tag{A23}$$

implying that

$$\left|P_0(iv) + P_1(iv)e^{-iv\tau_1}\right| = \left|P_2(iv) + P_3(iv)e^{-iv\tau_1}\right| \tag{A24}$$

or equivalently

$$\left(P_0 + P_1 e^{-iv\tau_1}\right)\left(\bar{P}_0 + \bar{P}_1 e^{iv\tau_1}\right) = \left(P_2 + P_3 e^{-iv\tau_1}\right)\left(\bar{P}_2 + \bar{P}_3 e^{iv\tau_1}\right)$$

where the arguments of P_0, P_1, P_2 and P_3 are omitted for the sake of simplicity, and overbar denotes complex conjugate. This equation can be rewritten as

$$|P_0|^2 + |P_1|^2 - |P_2|^2 - |P_3|^2 = 2A_1(v)\cos v\tau_1 - 2B_1(v)\sin v\tau_1 \tag{A25}$$

with

$$A_1(v) = \text{Re}\left(P_2\bar{P}_3 - P_0\bar{P}_1\right) \text{ and } B_1(v) = \text{Im}\left(P_2\bar{P}_3 - P_0\bar{P}_1\right).$$

Notice that the left hand side depends on only v, so this equation is a trigonometric equation for τ_1 with any fixed value of v. First, we check the existence of solutions. Two cases are now considered.

(A) The right hand side does not depend on τ_1, when $A_1(v) = B_1(v) = 0$ or $P_2\bar{P}_3 - P_0\bar{P}_1 = 0$. Let v be a solution of this equation. If $|P_0|^2 + |P_1|^2 - |P_2|^2 - |P_3|^2 \neq 0$, then there is no solution for τ_1, since (A25) is a contradiction. If $|P_0|^2 +$

$|P_1|^2 - |P_2|^2 - |P_3|^2 = 0$, then $\tau_1 > 0$ is arbitrary, and the corresponding values of τ_2 can be obtained from (A23),

$$\tau_2^m = \frac{1}{v}\left[\arg\left(-\frac{P_2 + P_3 e^{-iv\tau_1}}{P_0 + P_1 e^{-iv\tau_1}}\right) + 2m\pi\right] \quad \text{(A26)}$$

where m is an integer such that $\tau_2^m > 0$. If the denominator is zero, then from (A24), the numerator is also zero, so any $\tau_2 > 0$ is a solution.

(B) Assume that $[A_1(v)]^2 + [B_1(v)]^2 > 0$, then there exists $\phi_1(v)$ such that

$$\cos[\phi_1(v)] = \frac{A_1(v)}{\sqrt{[A_1(v)]^2 + [B_1(v)]^2}} \text{ and } \sin[\phi_1(v)] = \frac{B_1(v)}{\sqrt{[A_1(v)]^2 + [B_1(v)]^2}} \quad \text{(A27)}$$

or alternatively

$$\phi_1(v) = \arg\left[P_2(iv)\bar{P}_3(iv) - P_0(iv)\bar{P}_1(iv)\right].$$

Then, Eq. (A25) can be rewritten as follow:

$$|P_0|^2 + |P_1|^2 - |P_2|^2 - |P_3|^2 = 2\sqrt{[A_1(v)]^2 + [B_1(v)]^2}\cos[\phi_1(v) + v\tau_1] \quad \text{(A28)}$$

and by defining

$$\psi_1(v) = \cos^{-1}\left[\frac{|P_0|^2 + |P_1|^2 - |P_2|^2 - |P_3|^2}{2\sqrt{[A_1(v)]^2 + [B_1(v)]^2}}\right] \quad (\psi_1(v) \in [0, \pi]) \quad \text{(A29)}$$

we have

$$\tau_1^{n\pm} = \frac{1}{v}\left[\pm\psi_1(v) - \phi_1(v) + 2n\pi\right] \quad \text{(A30)}$$

where n is integer such that $\tau_1 > 0$. Notice that the sufficient and necessary condition for the existence of $\psi_1(v)$ is the inequality

$$\left||P_0|^2 + |P_1|^2 - |P_2|^2 - |P_3|^2\right| \le 2\sqrt{A_1^2 + B_1^2}. \quad \text{(A31)}$$

The solutions of this inequality give the crossing set Ω of all values of v which can result in stability switch. Since (A31) can be rewritten as a polynomial inequality, Ω consists of finitely many intervals Ω_k similarly to the previous case. For each $v \in \Omega$, the values of $\tau_1^{n\pm}$ are obtained from (A27), (A29) and (A30), and the corresponding values of τ_2 can be obtained from (A26):

$$\tau_2^m = \frac{1}{v}\left[\arg\left(-\frac{P_2(iv) + P_3(iv)e^{-iv\tau_1}}{P_0(iv) + P_1(iv)e^{-iv\tau_1}}\right) + 2m\pi\right] \quad \text{(A32)}$$

if the denominator is nonzero. If it is zero, then (A24) shows that the numerator is also zero, so any $\tau_2 > 0$ is solution.

Before processing further, some comments are in order.

1. In case (**A**) feasible value of v exists if and only if (A31) holds, which means that $v \in \Omega$.
2. Assume both the numerator and denominator of (A26) are equal to zero. Then,

$$\frac{P_0}{P_1} = \frac{P_2}{P_3}, \ |P_0| = |P_1| \text{ and } |P_2| = |P_3|$$

where the last two relations can be rewritten as

$$P_0 \bar{P}_0 = P_1 \bar{P}_1 \text{ and } P_2 \bar{P}_2 = P_3 \bar{P}_3.$$

Then,

$$\begin{aligned} P_1 \bar{P}_3 - P_0 \bar{P}_2 &= P_1 \frac{P_2 \bar{P}_2}{P_3} - P_0 \bar{P}_2 \\ &= \frac{\bar{P}_2 (P_1 P_2 - P_0 P_3)}{P_3} \\ &= 0. \end{aligned}$$

If $P_3 = 0$, then $P_2 = 0$ as well, so this relation holds obviously. Notice that this relation is the same as case (**A**) when τ_1 and τ_2 are interchanged.
3. By interchanging τ_1 and τ_2, we can proceed in the same way as it was shown for τ_1. It can be shown that

$$T_{n,m}^{k\pm} = \left\{ \left(\frac{\pm\psi_1(v) - \phi_1(v) + 2n\pi}{v}, \frac{\mp\psi_2(v) - \phi_2(v) + 2m\pi}{v} \right) \mid v \in \Omega_k \right\}$$

which is continuous in $\mathbb{R}^2$. Since at the left and right end points of Ω_k,

$$F\left(v_k^\ell\right) = F\left(v_k^r\right) = 0$$

with

$$F(v) = \left(|P_0|^2 + |P_1|^2 - |P_2|^2 - |P_3|^2\right)^2 - 4\left(A_1^2 + B_1^2\right),$$

clearly for $i = 1, 2$,

$$\cos\left[\psi_i(v_k^\ell)\right] = \delta_i^\ell \pi \text{ and } \cos\left[\psi_i(v_k^r)\right] = \delta_i^r \pi$$

where

$$\delta_i^\ell,\ \delta_i^r \in \{0; 1\}.$$

It can be also proved that

$$T_{n,m}^{k+} \text{ connects } T_{n+\delta_1^\ell,m-\delta_2^\ell}^{k-} \text{ and } T_{n+\delta_1^\ell,m-\delta_2^r}^{k-} \text{ at its two ends.}$$

Therefore, the stability switching curve for each Ω_k has the following possibilities:

- if $(\delta_1^\ell, \delta_2^\ell) = (\delta_1^r, \delta_2^r)$, then $T_{n,m}^{k+}$ and $T_{n+\delta_1^\ell,m-\delta_2^\ell}^{k-}$ form a loop on $\mathbb{R}^2$, so T is a set of closed continuous curves,
- if $(\delta_1^\ell, \delta_2^\ell) \neq (\delta_1^r, \delta_2^r)$, then T is a set of continuous curves with two endpoints either on the axes or extending to ∞ on the $\mathbb{R}_+^2$ region.

Theorem A.3 *The statement of Theorem A.2 remains valid, where now*

$$R_\ell = \text{Re}\left[P_\ell(iv)e^{-iv\tau_\ell} + P_3(iv)e^{-iv(\tau_1+\tau_2)}\right]$$

and

$$I_\ell = \text{Im}\left[P_\ell(iv)e^{-iv\tau_\ell} + P_3(iv)e^{-iv(\tau_1+\tau_2)}\right]$$

in the three-delay case.

The method described above consists of the following steps.

Step 1. Compute $A_1(v)$ and $B_1(v)$ and solve inequality (A31) to find the crossing v set, Ω.

Step 2. Let

$$\Omega_1 = \{v \mid v \in \Omega, A_1(v) = B_1(v) = 0\},$$

then for all $v \in \Omega_1$, any $\tau_1 > 0$ is solution.

Step 3. Let

$$\bar{\Omega}_1 = \Omega \backslash \Omega_1,$$

then for each $v \in \bar{\Omega}_1$, proceed as follows. Find first $\phi_1(v)$ from (A27), and then $\psi_1(v)$ by using (A29), and finally $\tau_1^{n\pm}$ from (A30). Now, for all $v \in \Omega$, we found all corresponding τ_1 values.

Step 4. For all (v, τ_1) pairs, the corresponding τ_2 values are given by (A32) if $P_0(iv) + P_1(iv)e^{-iv\tau_1} \neq 0$. Otherwise, all $\tau_2 > 0$ values are solutions.

Step 5. Find the directions of stability switches based on Theorem A.3.

Appendix B
Stability Analysis with Continuously Distributed Delays

Consider a linearized model with multiple delays

$$\dot{x}_\varepsilon(t) = \sum_{k=1}^{N} a_k x_\varepsilon(t - \tau_k)$$

where $x_\varepsilon(t)$ is the difference of $x(t)$ and the equilibrium level, and $\tau_1, \tau_2, \ldots, \tau_N$ are the delays. If the values of τ_k are uncertain, then they can be considered as random variables in the reasonable interval $[0, t]$. Assume that the density of τ_k is characterized by the gamma-type weighting function

$$W(t-s, m_k, T_k) = \begin{cases} \dfrac{1}{T_k} e^{-\frac{t-s}{T_k}} & \text{if } m_k = 0, \\[2ex] \dfrac{1}{m_k!}\left(\dfrac{m_k}{T_k}\right)^{m_k+1} (t-s)^{m_k} e^{-\frac{m_k(t-s)}{T_k}} & \text{if } m_k \geq 1, \end{cases} \tag{B1}$$

where m_k is a nonnegative integer and $T_k > 0$.

Then, by taking expectation on the right hand side, a Volterra-type integro-differential equation is obtained:

$$\dot{x}_\varepsilon(t) = \sum_{k=1}^{N} a_k \int_0^t W(t-s, m_k, T_k) x_\varepsilon(s) ds. \tag{B2}$$

Cushing (1977) offers a comprehensive analysis of this type of equations. In order to derive the characteristic equation, we look for the solution in exponential form

$$x_\varepsilon(t) = e^{\lambda t} u$$

and substitute it into the equation:

A. Matsumoto and F. Szidarovszky, *Dynamic Oligopolies with Time Delays*,
https://doi.org/10.1007/978-981-13-1786-6

$$\lambda e^{\lambda t} u = \sum_{k=1}^{N} a_k \int_0^t W(t-s, m_k, T_k) e^{\lambda s} u ds. \tag{B3}$$

Consider the general term of the right hand side and introduce the new variable $z = t - s$, when $s = t - z$ and $ds = -dz$:

$$\int_0^t W(z, m_k, T_k) e^{\lambda(t-z)} u dz = e^{\lambda t} \int_0^t W(z, m_k, T_k) e^{-\lambda z} u dz.$$

If $m_k = 0$, then the integral is the following:

$$\begin{aligned} \textstyle\int_0^t W(z, 0, T_k) e^{-\lambda z} dz &= \textstyle\int_0^t \frac{1}{T_k} e^{-\frac{z}{T_k}} e^{-\lambda z} dz \\ &= \textstyle\int_0^t \frac{1}{T_k} e^{-z\left(\frac{1}{T_k}+\lambda\right)} dz. \end{aligned}$$

Introduce the new variable

$$\bar{z} = z\left(\frac{1}{T_k} + \lambda\right),$$

assume that $\mathrm{Re}(\lambda) + \frac{1}{T_k} > 0$ and let $z \to \infty$. Then,

$$\int_0^\infty \frac{1}{T_k} e^{-\bar{z}} \frac{1}{\frac{1}{T_k} + \lambda} d\bar{z} = (1 + \lambda T_k)^{-1}. \tag{B4}$$

If $m_k \geq 1$, then similarly

$$\begin{aligned} \textstyle\int_0^t W(z, m_k, T_k) e^{-\lambda z} dz &= \textstyle\int_0^t \frac{1}{m_k!} \left(\frac{m_k}{T_k}\right)^{m_k+1} z^{m_k} e^{-\frac{m_k z}{T_k}} e^{-\lambda z} dz \\ &= \frac{1}{m_k!} \left(\frac{m_k}{T_k}\right)^{m_k+1} \textstyle\int_0^t z^{m_k} e^{-z\left(\frac{m_k}{T_k}+\lambda\right)} dz. \end{aligned}$$

By introducing the new variable

$$\bar{z} = z\left(\frac{m_k}{T_k} + \lambda\right),$$

and letting $z \to \infty$, we get the expression

$$\frac{1}{m_k!}\left(\frac{m_k}{T_k}\right)^{m_k+1}\int_0^\infty \frac{\bar{z}^{m_k}}{\left(\frac{m_k}{T_k}+\lambda\right)^{m_k}}e^{-\bar{z}}\frac{1}{\frac{m_k}{T_k}+\lambda}d\bar{z}$$

$$=\frac{1}{m_k!}\left(\frac{m_k}{T_k}\right)^{m_k+1}\int_0^\infty \frac{T_k^{m_k+1}}{(m_k+\lambda T_k)^{m_k+1}}\bar{z}^{m_k}e^{-\bar{z}}d\bar{z}$$

$$=\frac{1}{m_k!}\frac{m_k^{m_k+1}}{T_k^{m_k+1}}\frac{T_k^{m_k+1}}{m_k^{m_k+1}\left(1+\frac{\lambda T_k}{m_k}\right)^{m_k+1}}\int_0^\infty \bar{z}^{m_k}e^{-\bar{z}}d\bar{z} \tag{B5}$$

$$=\left(1+\frac{\lambda T_k}{m_k}\right)^{-(m_k+1)}$$

since the integral term is

$$\Gamma(m_k+1)=m_k!.$$

For the sake of simplicity, introduce the notation

$$q_k=\begin{cases}1 & \text{if } m_k=0,\\ m_k & \text{if } m_k\geq 1,\end{cases}$$

then Eq. (B3) can be reduced to equation

$$\lambda-\sum_{k=1}^{N}a_k\left(1+\frac{\lambda T_k}{q_k}\right)^{-(m_k+1)}=0 \tag{B6}$$

which can be rewritten as a polynomial equation for λ.

Consider next a polynomial equation

$$\lambda^n+a_1\lambda^{n-1}+a_2\lambda^{n-2}+\cdots++a_{n-1}\lambda+a_n=0. \tag{B7}$$

Introduce the Hurwitz matrices $n=1,2,3,\ldots$ as follows (Routh (1877), Hurwitz (1964)):

$$\boldsymbol{H}_1=(a_1),\ \boldsymbol{H}_2=\begin{pmatrix}a_1 & 1\\ 0 & a_2\end{pmatrix},\ \boldsymbol{H}_3=\begin{pmatrix}a_1 & 1 & 0\\ a_3 & a_2 & a_1\\ 0 & 0 & a_3\end{pmatrix},\ldots,$$

and in general

$$\boldsymbol{H}_n=\begin{pmatrix}a_1 & 1 & 0 & 0 & \cdots & 0\\ a_3 & a_2 & a_1 & 1 & \cdots & 0\\ a_5 & a_4 & a_3 & a_2 & \cdots & 0\\ \cdot & \cdot & \cdot & \cdot & \cdots & \cdot\\ 0 & 0 & 0 & 0 & \cdots & a_n\end{pmatrix}$$

where $a_j = 0$ for $j > n$. The famous Routh–Hurwitz criterion can be formulated as follows:

Theorem B.1 *All roots of equation (B7) are negative or have negative real parts if and only if the principal minors of the Hurwitz matrix* $\boldsymbol{H}_n$ *are positive.*

If $n = 1$, then $\det(\boldsymbol{H}_1) = a_1 > 0$ is the condition.
If $n = 2$, then $\det(\boldsymbol{H}_2) = a_1 a_2 > 0$.
If $n = 3$, then $\det(\boldsymbol{H}_3) = a_3(a_1 a_2 - a_3) > 0$.
If $n = 4$, then $\det(\boldsymbol{H}_4) = a_4\left(a_1 a_2 a_3 - a_1^2 a_4 - a_3^2\right) > 0$.

So if $n = 1$, then the only condition is that $a_1 > 0$. For larger values of n the following results can be proved. If $n = 2$, then in addition to this condition, $a_2 > 0$. If $n = 3$, then in addition to $a_1,\ a_2 > 0$, we need $a_3 > 0$ and $a_1 a_2 > a_3$. In the case of $n = 4$, in addition to $a_1 > 0,\ a_2 > 0,\ a_3 > 0$, we also need $a_4 > 0$ and $a_1 a_2 a_3 > a_1^2 a_4 + a_3^2$. Similar but more complicated conditions can be derived for $n \geq 5$.

The application of this methodology consists of two steps.

Step 1. Rewrite equation (B6) into a polynomial form.
Step 2. Use Theorem B.1 to check asymptotical stability or find conditions for asymptotic stability.

We also notice that Eq. (B2) can be rewritten as a system of ordinary differential equations in the following way. For the general term of the right hand side of Eq. (B2), introduce the following function where subscripts of m and T are omitted:

$$x_0(t) = \int_0^t \frac{m}{T} e^{-\frac{m(t-s)}{T}} x_\varepsilon(s) ds \tag{B8}$$

and

$$x_n(t) = \frac{1}{n!}\left(\frac{m}{T}\right)^{n+1} \int_0^t (t-s)^n e^{-\frac{m(t-s)}{T}} x_\varepsilon(s) ds \ (1 \leq n \leq m). \tag{B9}$$

Here, m and T are fixed; only the value of n changes. We know that

$$\frac{d}{dt}\int_0^t f(t,s)ds = \int_0^t \frac{d}{dt} f(t,s)ds + f(t,t),$$

so we have

$$\begin{aligned} \dot{x}_0(t) &= \textstyle\int_0^t -\left(\frac{m}{T}\right)^2 e^{-\frac{m(t-s)}{T}} x_\varepsilon(s)ds + \frac{m}{T} x_\varepsilon(t) \\ &= \frac{m}{T}\left(x_\varepsilon(t) - x_0(t)\right). \end{aligned} \tag{B10}$$

In general for $n \geq 1$,

$$\begin{aligned}
\dot{x}_n(t) &= \frac{d}{dt}\int_0^t \frac{1}{n!}\left(\frac{m}{T}\right)^{n+1}(t-s)^n e^{-\frac{m(t-s)}{T}} x_\varepsilon(s)ds \\
&= \frac{1}{n!}\left(\frac{m}{T}\right)^{n+1}\int_0^t \left[n(t-s)^{n-1}e^{-\frac{m(t-s)}{T}}x_\varepsilon(s) - \frac{m}{T}(t-s)^n e^{-\frac{m(t-s)}{T}}x_\varepsilon(s)\right]ds \\
&= \frac{m}{T}\left\{\frac{1}{(n-1)!}\left(\frac{m}{T}\right)^n\int_0^t (t-s)^{n-1}e^{-\frac{m(t-s)}{T}}x_\varepsilon(s)ds - \frac{1}{n!}\left(\frac{m}{T}\right)^{n+1}\int_0^t (t-s)^n e^{-\frac{m(t-s)}{T}}x_\varepsilon(s)ds\right\} \\
&= \frac{m}{T}\left(x_{n-1}(t) - x_n(t)\right)
\end{aligned} \tag{B11}$$

Based on Eqs. (B8)–(B11), the Volterra-type integro-differential equation can be rewritten as a system of $\sum_{k=1}^{N} m_k + (N+1)$ ordinary differential equations

$$\dot{x}_\varepsilon(t) = \sum_{k=1}^{N} a_k x_{km_k}(t)$$

$$\left.\begin{aligned}
\dot{x}_{k0}(t) &= \frac{m_k}{T_k}\left(x_\varepsilon(t) - x_{k0}(t)\right) \\
\dot{x}_{kn}(t) &= \frac{m_k}{T_k}\left(x_{k,n-1}(t) - x_{kn}(t)\right)\ (1 \le n \le m_k)
\end{aligned}\right\}(1 \le k \le N) \tag{B12}$$

with the initial conditions $x_\varepsilon(0) =$ given, $x_{kn}(0) = 0\ (0 \le n \le m_k)$. So the numerical solution of equation (B2) can be done by solving the differential equations (B12) by any standard method. For example, an accurate Runge–Kutta method is applicable (see, for example, Szidarovszky and Yakowitz 1986).

In the weighting function (B1), m_k is a nonnegative integer.

(a) If $m_k = 0$, then weights are exponentially declining with the most weight given to the most current value of $x(t)$.
(b) If $m_k \ge 1$, then zero weight is assigned to the most current $x(t)$ value, and the weight increases until its maximum point at $t - s = T_k$ and then decreases exponentially thereafter.
(c) The area under the weighting function in interval $[0, \infty)$ is unity.
(d) As m_k increases, the weighting function becomes more peaked around $t - s = T_k$, and for sufficiently large values of m_k, the function may, for all practical purposes, be regarded as very close to the Dirac delta function centered at $t - s = T_k$.
(e) As $T_k \to 0$, the weighting function tends to the Dirac delta function centered at zero.

Property (a) is obvious, since at $m_k = 0$ the weighting function is a decreasing exponential function. Property (b) follows from the simple fact that with notation $z = t - s$,

$$W(z, m, T) = \frac{1}{m!}\left(\frac{m}{T}\right)^{m+1} z^m e^{-\frac{mz}{T}}$$

and

$$\frac{\partial W(z,m,T)}{\partial z} = \frac{1}{m!}\left(\frac{m}{T}\right)^{m+1}\left\{mz^{m-1}e^{-\frac{mz}{T}} - \frac{m}{T}z^m e^{-\frac{mz}{T}}\right\}$$

$$= \frac{1}{m!}\left(\frac{m}{T}\right)^{m+1}\frac{m}{T}z^{m-1}e^{-\frac{mz}{T}}\,(T-z)\,.$$

Property (c) is a simple consequence of the definition of the gamma function, since

$$\int_0^\infty \frac{1}{m!}\left(\frac{m}{T}\right)^{m+1} z^m e^{-\frac{mz}{T}}\,dz = \int_0^\infty \frac{1}{m!}\left(\frac{m}{T}\right)^{m+1}\left(\frac{T}{m}\right)^m t^m e^{-t}\frac{T}{m}dt$$

$$= \frac{1}{m!}\Gamma(m+1) = 1$$

where the new integration variable $t = \frac{mz}{T}$ is introduced.

In the case of property d), notice first that from the Stirling formula

$$m! \simeq \sqrt{2\pi m}\left(\frac{m}{e}\right)^m \quad \text{for large values of } m,$$

so with the notation of $z = t - s$,

$$W(z,m,T) \sim \frac{e^m m^{m+1}}{\sqrt{2\pi m} m^m T^{m+1}} z^m e^{-\frac{mz}{T}} = \frac{\left(\frac{z}{T}\right)^m m^{\frac{1}{2}}}{\sqrt{2\pi} T e^{m(\frac{z}{T}-1)}}.$$

If $z = T$, then

$$W(T,m,T) \simeq \frac{m^{\frac{1}{2}}}{\sqrt{2\pi}T} \to \infty \text{ as } m \to \infty.$$

If $z \neq T$, then with notation $\alpha = \frac{z}{T}$,

$$W(z,m,T) \simeq \frac{m^{\frac{1}{2}}}{\sqrt{2\pi}T}\left(\frac{\alpha}{e^{\alpha-1}}\right)^m \to 0 \text{ as } m \to \infty,$$

since with any $\alpha > 0$, $\alpha < e^{\alpha-1}$, which is implied by the well know relation $e^x > x+1$ for all nonzero real values of x.

In showing property (e), notice first that

$$W(T,m,T) = \frac{1}{m!}\left(\frac{m}{T}\right)^{m+1} T^m e^{-m} = \frac{m^{m+1}e^{-m}}{m!T} \to \infty \text{ as } T \to 0,$$

and with any fixed $z = t - s$,

$$W(z,m,t) = \frac{1}{m!}\left(\frac{m}{T}\right)^{m+1} z^m e^{-\frac{mz}{T}} \to 0 \text{ as } T \to 0$$

since

$$T^{m+1}e^{\frac{mz}{T}} \to \infty.$$

Observe now that from property (d), we can see that with large values of m,

$$\int_0^t W(t-s,m,T)x(s)ds = \int_0^t W(z,m,T)x(t-z)dz \simeq x(t-T)$$

which is the case of fixed delay T. From property e), it is also clear that with small values of T,

$$\int_0^t W(t-s,m,T)x(s)ds = \int_0^t W(z,m,T)x(t-z)dz \simeq x(t),$$

so we can recover the case where no delay is present. Based on these two observations, we can consider the case of continuously distributed delays as a "transition" from the no-delay case to fixed delay.

Appendix C
Comparison of Discrete and Continuous Dynamics

Consider an n-dimensional continuous system:

$$\dot{x}_i(t) = f_i\left(x_1(t), \ldots, x_n(t)\right) \text{ for } i = 1, 2, \ldots, n \tag{C1}$$

which can be written in vector form as

$$\dot{\boldsymbol{x}}(t) = \boldsymbol{f}\left(\boldsymbol{x}(t)\right) \tag{C2}$$

with

$$\boldsymbol{x}(t) = \left(x_1(t), \ldots, x_n(t)\right) \text{ and } \boldsymbol{f} = \left(f_1, f_2, \ldots, f_n\right).$$

The Jacobian of system (C1) is the $n \times n$ matrix

$$\boldsymbol{J}_C(\boldsymbol{x}) = \begin{pmatrix} \dfrac{\partial f_1(\boldsymbol{x})}{\partial x_1} & \cdots & \dfrac{\partial f_1(\boldsymbol{x})}{\partial x_n} \\ \cdot & \cdots & \cdot \\ \cdot & \cdots & \cdot \\ \cdot & \cdots & \cdot \\ \dfrac{\partial f_n(\boldsymbol{x})}{\partial x_1} & \cdots & \dfrac{\partial f_n(\boldsymbol{x})}{\partial x_n} \end{pmatrix} \tag{C3}$$

where the rows are the gradient vectors of functions $f_1(\boldsymbol{x}), \ldots, f_n(\boldsymbol{x})$, respectively. Let $\boldsymbol{x}^*$ be a steady state of the system, that is, $\boldsymbol{f}\left(\boldsymbol{x}^*\right) = \boldsymbol{0}$. It is well known that steady state $\boldsymbol{x}^*$ is locally asymptotically stable if all eigenvalues of $\boldsymbol{J}_C(\boldsymbol{x}^*)$ are negative or have negative real parts, and if at least one eigenvalue is positive or have positive real part, then $\boldsymbol{x}^*$ is locally unstable.

The corresponding discrete system is obtained by replacing the derivative $\dot{x}_i(t)$ by $x_i(t+1) - x_i(t)$ to get

$$x_i(t+1) = x_i(t) + f_i\left(x_1(t), \ldots, x_n(t)\right) \text{ for } i = 1, 2, \ldots, n \tag{C4}$$

A. Matsumoto and F. Szidarovszky, *Dynamic Oligopolies with Time Delays*,
https://doi.org/10.1007/978-981-13-1786-6

or in matrix form

$$\boldsymbol{x}(t+1) = \boldsymbol{x}(t) + \boldsymbol{f}(\boldsymbol{x}(t)). \tag{C5}$$

The Jacobian of this system is clearly

$$\boldsymbol{J}_D(\boldsymbol{x}) = \boldsymbol{I} + \boldsymbol{J}_C(\boldsymbol{x}) \tag{C6}$$

where $\boldsymbol{I}$ is the $n \times n$ identity matrix.

Clearly, systems (C2) and (C5) have the same steady state. It is well known that $\boldsymbol{x}^*$ is locally asymptotically stable with respect to dynamic system (C5) if all eigenvalues of $\boldsymbol{J}_D(\boldsymbol{x}^*)$ are inside the unit circle, and if at least one eigenvalue is outside the unit circle, then $\boldsymbol{x}^*$ is locally unstable.

In comparing (C3) and (C6), we get the following observations:

(a) If λ_i $(i = 1, 2, \ldots, n)$ are the eigenvalues of $\boldsymbol{J}_C$, then $\lambda_i + 1$ $(i = 1, 2, \ldots, n)$ are the eigenvalues of $\boldsymbol{J}_D$.
(b) If λ_i $(i = 1, 2, \ldots, n)$ are the eigenvalues of $\boldsymbol{J}_D$, then $\lambda_i - 1$ $(i = 1, 2, \ldots, n)$ are the eigenvalues of $\boldsymbol{J}_C$.

These two facts imply the following result.

Lemma C.1 *Let λ_i^D and $\lambda_i^C = \lambda_i^D - 1$ be a pair of eigenvalues of $\boldsymbol{J}_C$ and $\boldsymbol{J}_D$. Then,*

(i) If $|\lambda_i^D| < 1$, then $\mathrm{Re}\,\lambda_i^C < 0$,
(ii) If $\mathrm{Re}\,\lambda_i^C < 0$, then $|\lambda_i^D| < 1$ is not necessarily true.

Proof (i) If $|\lambda_i^D| < 1$, then λ_i^D is inside a circle with center 0 and unit radius. Since $\lambda_i^C = \lambda_i^D - 1$, λ_i^C is inside a circle with center -1 and unit radius. As Fig. C.1 shows, this circle is inside the left half of the complex plane.

(ii) If $\mathrm{Re}\,\lambda_i^C < 0$, then $\mathrm{Re}\,\lambda_i^D < 1$, and since $\mathrm{Im}\,\lambda_i^C$ can be arbitrary, there is no guarantee that $\left(\mathrm{Re}\,\lambda_i^D\right)^2 + \left(\mathrm{Im}\,\lambda_i^D\right)^2 < 1$. Notice that the region of the eigenvalues such that $\mathrm{Re}\,\lambda_i^D < 1$ is left of the $\mathrm{Re}\,\lambda = 1$ vertical line in the complex plane and the unit circle is a small subset of it as shown in Fig. C.2. ■

The eigenvalues are the roots of the polynomial equations

$$\det\left(\boldsymbol{J}_C(\boldsymbol{x}^*) - \lambda\boldsymbol{I}\right) = 0 \text{ or } \det\left(\boldsymbol{J}_D(\boldsymbol{x}^*) - \lambda\boldsymbol{I}\right) = 0$$

which have degree n. Let

$$p^C(\lambda) = \lambda^n + a_1\lambda^{n-1} + a_2\lambda^{n-2} + \cdots + a_{n-1}\lambda + a_n \tag{C7}$$

be the characteristic polynomial of $\boldsymbol{J}_C$, then clearly

$$\begin{aligned} p^D(\lambda) &= (\lambda-1)^n + a_1(\lambda-1)^{n-1} + a_2(\lambda-1)^{n-2} + \cdots + a_{n-1}(\lambda-1) + a_n \\ &= \lambda^n + b_1\lambda^{n-1} + b_2\lambda^{n-2} + \cdots + b_{n-1}\lambda + b_n \end{aligned} \tag{C8}$$

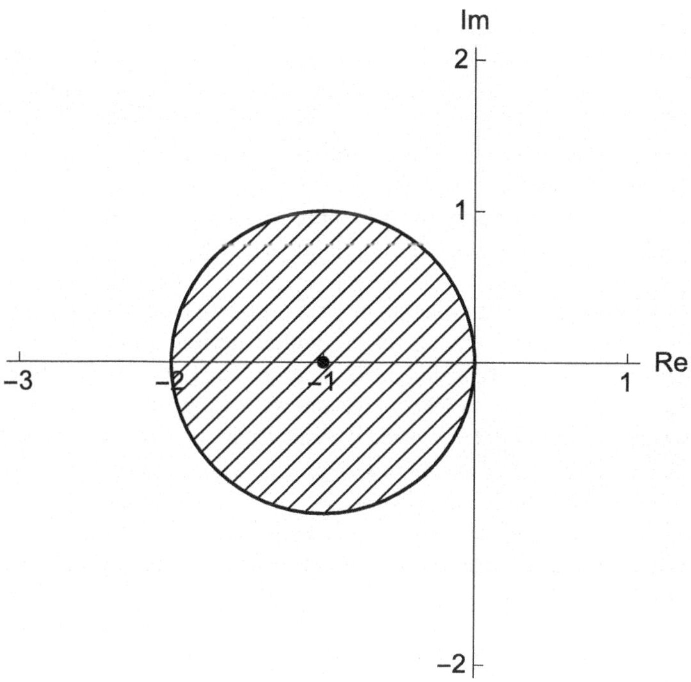

Fig. C.1 Illustration of Proof (i) of Lemma C.1

Fig. C.2 Illustration of Proof (ii) of Lemma C.1

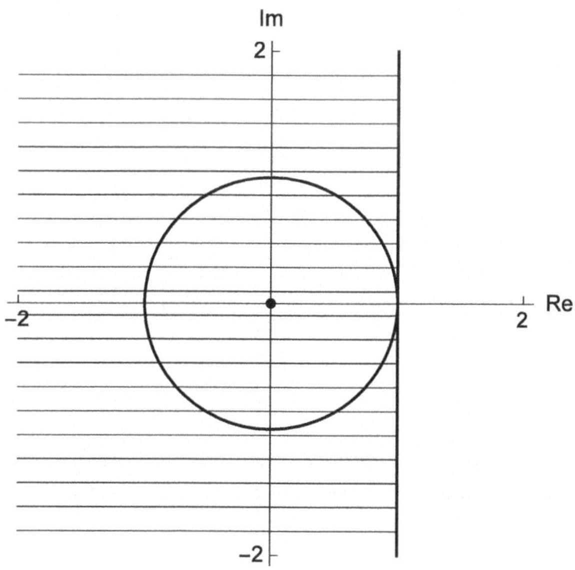

is the characteristic polynomial of $\boldsymbol{J}_D$.

The Routh–Hurwitz criterion gives the sufficient and necessary conditions that all root of $p^C(\lambda)$ are negative or have negative real parts. It is discussed in Appendix B:

In the case of $n = 1$, this condition is $a_1 > 0$.
If $n = 2$, then $a_1,\ a_2 > 0$.
If $n = 3$, then $a_1,\ a_2,\ a_3 > 0$ and $a_1 a_2 > a_3$
If $n = 4$, then $a_1,\ a_2,\ a_3,\ a_4 > 0$ and $a_1 a_2 a_3 > a_1^2 a_4 + a_3^2$.

In more general cases $n > 4$, similar but more complicated conditions can be derived.

In the discrete case, however, there is no simple criterion to guarantee that all roots of an n-degree polynomial are inside the unit circle. The Schur–Cohn algorithm (see for example, Stoica and Moses 1992) gives, however, a general method to check if all roots of a given polynomial are inside the unit circle, which cannot be represented by easy and simple rules. In cases of low-degree polynomials, the conditions can be presented in simple inequalities.

If $n = 1$, then the condition is $|b_1| < 1$ or equivalently $-1 < b_1 < 1$.

In the case of $n = 2$, the conditions are as follows (see Bischi et al. 2010):

$$\begin{gathered} b_2 < 1 \\ \pm b_1 + b_2 + 1 > 0. \end{gathered} \tag{C9}$$

When $n = 3$, the conditions become much more complex (Farebrother 1973):

$$\begin{gathered} 1 + b_1 + b_2 + b_3 > 0, \\ 1 - b_1 + b_2 - b_3 > 0, \\ 1 - b_2 + b_1 b_3 - b_3^2 > 0, \\ 3 - b_2 > 0. \end{gathered} \tag{C10}$$

In the case of $n = 4$, even more complicated conditions can be derived (Farebrother 1973):

$$\begin{gathered} 1 - b_4 > 0, \\ 3 + 3b_4 > b_2, \\ 1 + b_1 + b_2 + b_3 + b_4 > 0, \\ 1 - b_1 + b_2 - b_3 + b_4 > 0, \\ (1 - b_4)\left(1 - b_4^2\right) - b_2 (1 - b_4)^2 + (b_1 - b_3)(b_3 - b_1 b_4) > 0. \end{gathered} \tag{C11}$$

Matsumoto and Szidarovszky (2015a) offer additional details, computer studies and particular applications to economic dynamics.

References

Agiza, H. N., & Elsadany, A. (2003). Nonlinear dynamics in the Cournot duopoly game with heterogeneous players. *Physica A*, *320*(1–4), 512–524.

Al-Nowaihi, A., & Levine, P. L. (1985). The stability of the Cournot oligopoly model: A reassessment. *Journal of Economic Theory*, *35*, 307–321.

Bellman, R. E., & Cooke, K. L. (1963). *Differential-difference equations*. New York: Academic Press.

Bischi, G.-I., Chiarella, C., Kopel, M., & Szidarovszky, F. (2010). *Nonlinear oligopolies: Stability and bifurcations*. Berlin: Springer.

Burger, E. (1959). *Einfuhrung in die Theorie der Spiele*. Berlin: De Gruyter.

Cheng, L. (1985). Comparing Bertrand and Cournot equilibria: A geometric approach. *Rand Journal of Economics*, *16*, 146–152.

Chiarella, C., & Khomin, A. (1996). An analysis of the complex dynamic behavior of nonlinear oligopoly models with time lags. *Chaos, Solitons and Fractals*, *7*(12), 2049–2065.

Chiarella, C., Matsumoto, A., & Szidarovszky, F. (2013). Isoelastic oligopolies under uncertainty. *Applied Mathematics and Computation*, *219*, 10475–10486.

Chiarella, C., & Szidarovszky, F. (2004). Dynamic oligopolies without full information and with continuously distributed time lags. *Journal of Economic Behavior and Organization*, *54*(4), 495–511.

Chiarella, C., & Szidarovszky, F. (2001). The nonlinear Cournot model under uncertainty with continuously distributed time lags. *Central European Journal of Operations Research*, *9*(3), 183–196.

Cournot, A. (1838). *Recherches sur les principes mathématiques de la théorie des richessess*. Hachette, Paris (English translation (1960) *Researches into the mathematical principles of the theory of wealth*, Kelley, New York).

Cox, J. C., & Walker, M. (1998). Learning to play Cournot duopoly strategies. *Journal of Economic Behavior and Organization*, *36*, 141–161.

Cushing, J. M. (1977). *Integro-differential equations and delay models in population dynamics*. Berlin: Springer.

Cyert, R. M., & DeGroot, M. H. (1973). An analysis of cooperation and learning in a duopoly context. *American Economic Review*, *63*(1), 24–37.

Cyert, R. M., & DeGroot, M. H. (1987). *Bayesian analysis and uncertainty in economic theory*. Totowa: Rowman & Littlefield.

DeGroot, M. H. (1970). *Optimal statistical decisions*. New York: McGraw Hill.

A. Matsumoto and F. Szidarovszky, *Dynamic Oligopolies with Time Delays*,
https://doi.org/10.1007/978-981-13-1786-6

Dixit, A. (1979). A model of duopoly suggesting a theory of entry barriers. *Bell Journal of Economics, 10*, 20–32.

Elsadany, A., & Matouk, A. (2014). Dynamic Cournot duopoly game with delay. *Journal of Complex Systems*,. https://doi.org/10.1155/2014/384843.

Farebrother, R. (1973). Simplified Samuelson conditions for cubic and quartic equations. *The Manchester School of Economic and Social Studies, 41*, 395–400.

Fisher, F. M. (1961). The stability of the Cournot solution: The effects of speeds of adjustment and increasing marginal costs. *Review of Economic Studies, 28*, 125–135.

Flath, D. (1992). Horizontal shareholding interlocks. *Managerial and Decision Economics, 13*, 75–77.

Flath, D. (1991). When is it rational for firms to acquire silent interests in rivals? *International Journal of Industrial Organization, 9*, 573–583.

Friedman, J. W. (1981). Oligopoly theory. In K. J. Arrow & M. D. Intriligator (Eds.), *Handbook of mathematical economics*. Oxford: Oxford University Press.

Friedman, J. W. (1977). *The Theory of game and oligopoly*. Amsterdam: North Holland Publishing Company.

Fudenberg, D., & Levine, D. (1998). *The theory of learning in games*. Cambridge, MA: MIT Press.

Furth, D. (1986). Stability and instability in oligopoly. *Journal of Economic Theory, 40*, 197–228.

Gandolfo, G. (2009). *Economic dynamics* (4th ed.). Berlin: Springer.

Gu, K., Nicolescu, S.-I., & Chen, J. (2005). On stability crossing curves for general systems with two delays. *Journal of Mathematical Analysis and Applications, 311*, 231–253.

Häckner, J. (2000). A note on price and quantity competition in differentiated oligopolies. *Journal of Economic Theory, 93*, 233–239.

Hadar, J. (1966). Stability of oligopoly with product differentiation. *Review of Economic Studies, 33*, 57–60.

Hahn, F. (1962). The stability of the Cournot oligopoly solution. *Review of Economic Studies, 29*, 329–331.

Hale, J. K., & Kocak, H. (1991). *Dynamics and bifurcations*. Texts in Applied Mathematics (Vol. 3). New York: Springer.

Hurwitz, A. (1964). On the conditions under which an equation has only roots with negative real parts. In R. Bellman & R. E. Kalaba (Eds.), *Selected papers on mathematical trends in control theory*. New York: Dover.

Invernizzi, S., & Medio, A. (1991). On lags and chaos in economic dynamic models. *Journal of Mathematical Economics, 20*, 521–550.

Kopel, M. (1996). Simple and complex adjustment dynamics in Cournot models. *Chaos, Solitons and Fractals, 7*(12), 2031–2048.

Kuang, Y. (1993). *Delay differential equations with applications in population dynamics*. Boston: Academic Press.

Kubicek, M., & Marek, M. (1986). *Computational methods in bifurcation theory and dissipative structures*. Berlin: Springer.

Kuznetsov, Y. A., & Sacker, R. J. (2008). Neimark-Sacker bifurcation. *Scholarpedia, 3*(5), 1845.

Leonard, D., & Nishimura, K. (1999). Nonlinear dynamics in the Cournot model without full information. *Annals of Operations Research, 89*, 165–173.

Lin, X., & Wang, H. (2012). Stability analysis of delay differential equations with two discrete delays. *Canadian Applied Mathematics Quarterly, 20*(4), 519–532.

Martelli, M. (1999). *Introduction to discrete dynamical systems and chaos*. New York: Wiley-Interscience.

Matsumoto, A., Qiao, H., & Szidarovszky, F. (2012a). Learning in dynamic oligopolies with time delay. In *Presented in the 2012 China International Conference on Game Theory and Applications, August 29–30, 2012, Qingdao, China*.

Matsumoto, A., Merlone, U., & Szidarovszky, F. (2012b). Some notes on applying the Herfindahl-Hirschman index. *Applied Economic Letters, 19*, 181–184.

Matsumoto, A., Merlone, U., & Szidarovszky, F. (2010a). Dynamic oligopoly with partial cooperation and antitrust threshold. *Journal of Economic Behavior and Organization*, *73*, 259–272.

Matsumoto, A., Merlone, U., & Szidarovszky, F. (2010). Cartelizing groups in dynamic hyperbolic oligopoly with antitrust threshold. *Australian Economic Papers*, *2010*, 289–300.

Matsumoto, A., Merlone, U., & Szidarovszky, F. (2008). Cartelizing groups in dynamic linear oligopoly with antitrust threshold. *International Game Theory Review*, *10*(4), 399–419.

Matsumoto, A., Qiao, H., & Szidarovszky, F. (2014). Adaptive learning in dynamic oligopolies with time delay. In L. A. Petrosyan, V. Romanovsky, & D. W. Yeung (Eds.), *Advances in economics and optimization* (pp. 71–89). Hauppauge: Nova Publisher.

Matsumoto, A., & Szidarovszky, F. (2016). Learning in monopolies with delayed price information. In A. Matsumoto, F. Szidarovszky, & T. Asada (Eds.), *Essays in economic dynamics* (pp. 57–80). Singapore: Springer Science + Business Media.

Matsumoto, A., & Szidarovszky, F. (2015a). *On the comparison of discrete and continuous dynamic systems* (p. 248). Tokyo, Japan, Institute of Economic Research, Discussion paper No: Chuo University.

Matsumoto, A., & Szidarovszky, F. (2015b). Dynamic monopoly with multiple continuously distributed time delays. *Mathematics and Computers in Simulations*, *108*, 99–118.

Matsumoto, A., & Szidarovszky, F. (2015c). Delayed dynamics of a Cournot game with heterogeneous duopolies. *Applied Mathematics and Computations*, *269*, 699–713.

Matsumoto, A., & Szidarovszky, F. (2015d). Nonlinear Cournot duopoly with implementation delays. *Chaos, Solitons and Fractals*, *79*, 157–165.

Matsumoto, A., & Szidarovszky, F. (2015e). *Game theory and its applications*. Tokyo: Springer.

Matsumoto, A., & Szidarovszky, F. (2015f). Learning in monopolies with delayed feedback on price expectations. *Communications in Nonlinear Science and Numerical Simulation*, *28*, 151–165.

Matsumoto, A., & Szidarovszky, F. (2015g). Dynamic economic systems with two time delays. *International Journal of Economic Behavior and Organization*, *3*(2), 77–85.

Matsumoto, A., & Szidarovszky, F. (2014a). Discrete-time delay dynamics of boundedly rational monopoly. *Decisions in Economics and Finance*, *37*, 53–79.

Matsumoto, A., & Szidarovszky, F. (2014b). Boundedly rational monopoly with single continuously distributed time delay. In R. Dieci, X-Z. He, C. Hommes (Eds.), *Nonlinear economic dynamics and financial modeling* (pp. 83–107). Heidelberg: Springer.

Matsumoto, A., & Szidarovszky, F. (2014c). Discrete and continuous dynamics in nonlinear monopolies. *Applied Mathematics and Computations*, *232*, 632–642.

Matsumoto, A., & Szidarovszky, F. (2014d). Complex dynamics of monopolies with gradient adjustment. *Economic Modeling*, *42*, 220–229.

Matsumoto, A., & Szidarovszky, F. (2014e). Theocharis problem reconsidered in differentiated oligopoly. *Economic Research International*. https://doi.org/10.1155/2014/630351.

Matsumoto, A., & Szidarovszky, F. (2014). Dynamic monopoly with demand delay. *Theoretical Economic Letters*, *2014*(4), 146–154.

Matsumoto, A., Chiarella, C., & Szidarovszky, F. (2013). Dynamic monopoly with bounded continuously distributed delay. *Chaos, Solitons and Fractals*, *47*, 66–72.

Matsumoto, A., & Szidarovszky, F. (2012a). Nonlinear delay monopoly with bounded rationality. *Chaos, Solitons and Fractals*, *45*, 507–519.

Matsumoto, A., & Szidarovszky, F. (2012b). Dynamics of boundedly rational monopoly with time delay. *The Journal of Economics (Chuo University, Tokyo)*, *52*(4), 105–121.

Matsumoto, A., & Szidarovszky, F. (2011a). Mixed Cournot-Bertrand competition in N-firm differentiated oligopolies, Unpublished manuscript, Department of Economics, Chuo University, Tokyo.

Matsumoto, A., & Szidarovszky, F. (2011b). Price and quantity competition in dynamic oligopolies with product differentiation. *Revista Investigation Operacional*, *32*(3), 204–219.

Matsumoto, A., & Szidarovszky, F. (2011c). Stability, bifurcation and chaos in N-firm nonlinear Cournot games. *Discrete Dynamics in Nature and Society*. https://doi.org/10.1155/2011/380530.

Matsumoto, A., Szidarovszky, F., & Yoshida, H. (2011). Dynamics in linear Cournot oligopolies with two time delays. In G-I. Bischi, C. Chiarella, & L. Gardini (Eds.), *Computational Economics* (pp. 311–327). Berlin: Springer.

Matsumoto, A., & Szidarovszky, F. (2010). Delay dynamics in heterogeneous competition with product differentiation. *Nonlinear Analysis: Real World Applications, 11*, 601–611.

Matsumoto, A., & Szidarovszky, F. (2007). Delayed nonlinear Cournot and Bertrand dynamics with product differentiation. *Nonlinear Dynamics, Psychology, and Life Sciences, 11*(3), 367–395.

Matsumoto, A., & Onozaki, T. (2005). Heterogeneous strategies in nonlinear duopoly with product differentiation. *Pure Mathematics and Applications, 16*, 443–466.

Merlone, U. (2007). Shareholding interlocks: Profit formulations and cartelizing effects. *Journal of Interdisciplinary Mathematics, 10*(1), 99–117.

Nwogugu, M. (2005). Some new anti-trust model. *SSRN eLibray.*

Nyarko, Y. (1991). Learning in mis-specified models and the possibility of cycles. *Journal of Economic Theory, 55*, 416–427.

Okuguchi, K. (1984). Equilibrium prices in the Bertrand and Cournot oligopolies. *Journal of Economic Theory, 42*, 128–139.

Okuguchi, K. (1976). *Expectations and stability in oligopoly models.* Berlin: Springer.

Okuguchi, K., & Szidarovszky, F. (1999). *The theory of oligopoly with multi-product firms* (2nd ed.). Berlin: Springer.

Porter, R. H., & Zona, J. D. (1999). Ohio school milk markets: An analysis of bidding. *The Rand Journal of Economics, 30*, 263–288.

Puu, T. (2003). *Attractors, bifurcations, and chaos: Nonlinear phenomena in economics* (2nd ed.). Berlin: Springer.

Puu, T. (1991). Chaos in business cycles. *Chaos, Solitons and Fractals, 1*(5), 457–473.

Puu, T., & Sushko, I. (Eds.). (2002). *Oligopoly dynamics.* Berlin: Springer.

Reynolds, R. J., & Snapp, B. R. (1986). The competitive effects of partial equity interest and joint ventures. *International Journal of Industrial Organization, 4*, 141–153.

Reynolds, R. J., & Snapp, B. R. (1982). The economic effect of partial equity interests and joint ventures. US Department of Justice, Antitrust Division, Economic Policy Office, Discussion Paper #3.

Routh, E. J. (1877). *A treatise on the stability of a given state of motion, particularly steady motion.* London: Macmillan and Co.

Singh, N., & Vives, X. (1984). Price and quantity competition in a differentiated duopoly. *Rand Journal of Economics, 15*, 546–554.

Stoica, P., & Moses, R. L. (1992). On the unit circle problem: The Schur-Cohn procedure revisited. *Signal Processing, 26*, 95–118.

Szep, J., & Forgo, F. (1985). *Introduction to the theory of games.* Budapest: Akadémiai Kiado.

Szidarovszky, F., & Krawczyk, J. B. (2004). On stable learning in dynamic oligopolies. *Pure Mathematics and Applications, 15*(4), 453–468.

Szidarovszky, F., & Bahill, T. (1998). *Linear systems theory* (2nd ed.). Boca Raton: CRC Press.

Szidarovszky, F., & Yakowitz, S. (1986). *Principles and procedures of numerical analysis.* New York: Plenum Press.

Theocharis, R. D. (1959). On the stability of the Cournot solution on the oligopoly problem. *Review of Economic Studies, 27*, 133–134.

Viscusi, W. K., Harrington, J. E., & Vernon, J. M. (2005). *Economics of regulation and antitrust* (4th ed.). Cambridge, MA: MIT Press.

Winston, M. D. (2006). *Lectures on antitrust economics.* Cambridge, MA: MIT Press.

Young, H. P. (2004). *Strategic learning and its limits.* Oxford: Oxford University Press.

Yousefi, S., & Szidarovszky, F. (2005). Once more on price and quantity competition in differentiated duopolies: A simulation study. *Pure Mathematics and Applications, 16*, 515–531.

Index

A. Matsumoto and F. Szidarovszky, *Dynamic Oligopolies with Time Delays*,
https://doi.org/10.1007/978-981-13-1786-6

The manufacturer's authorised representative in the EU is Springer Nature Customer Service Centre GmbH, Europaplatz 3, 69115 Heidelberg, Germany. If you have any concerns regarding our products, please contact ProductSafety@springernature.com

Printed and bound by CPI Group (UK) Ltd, Croydon, CR0 4YY
25/09/2026
02222900-0010